Penny Ante Science
6th Edition

Cynthia E. Ledbetter, Ph.D.
Fred L. Fifer, Jr., Ph.D.

Penny Ante Science

Copyright 2016 Near-Normal Design and Production Studio, all rights reserved. *Penny Ante Science*© activities may be copied by teachers for use in their classrooms only. For use other than that described, contact the authors in writing at Near-Normal Design and Production Studio, 1540 Keller Pkwy #108-261, Keller, TX 76248. Be sure to visit our web site at www.nearnormalstudio.com or Near-Normal Design and Production Studio on Facebook for updates and more information.

Penny Ante Science is a trademark of Near-Normal Design and Production Studio. We are the sole source of the *Penny Ante Science* activity books and *NOW* class starter books. Near-Normal Design and Production Studio is owned by Cynthia E. Ledbetter. Our books may be purchased on demand at Amazon.com.

Original art is by Cynthia E. Ledbetter and Fred L. Fifer. Clip art is from Print Artist, Creative Commons and Microsoft.

ISBN-13: 978-1532764592
ISBN-10: 1532764596

Notes to Teachers

Long ago Fred Fifer and Cynthia Ledbetter were faced with the issue of teaching hands-on science with little or no equipment and no budget with which to purchase supplies. Independently, they began writing their own labs, adapting what was in textbooks and building, borrowing, begging, or buying what they could to be able to teach science. Years later they met at The University of Texas at Dallas and the *Penny Ante Science* series was born. The activities in this book are designed to be used with any type of student. The authors have resisted the temptation to assign grade levels because a lesson that appears appropriate for young children may be used as a lesson introduction for students in advanced classes. More complicated lessons can be simplified for less sophisticated children; to do this you may have to change the difficulty of questioning, or read the directions to the children. The authors encourage you to adapt these activities as you see fit, but to ensure that they remain discovery rather than confirmatory type activities.

While activities are divided into the traditional science sections, they are also indexed in a manner that allows you to integrate the sciences. For instance, you may use *Scents or Sense* as a means to teach observation skills, physical properties of matter, or as a focus for a study of the nervous system activity. *The Great Buffalo Shortage* may be an earth science, an environmental science, or a physics activity.

Several of the activities in this book use potentially dangerous equipment (hot plates, boiling water, open flames, etc.). Safety equipment is listed and directions for use are given. **Please explain all safety procedures to your students prior to doing the activities; require them to follow these procedures. Please review and follow all safety procedures prior to doing any preparation required of the teacher**. The authors assume no responsibility for accidents that may occur during performance of these activities.

Drs. Fifer and Ledbetter are former public school science teachers with several hundred hours of experience training elementary, middle school and high school teachers. We hope you and your students enjoy these activities. If you have any questions or comments, please contact us at Near-Normal Design and Production Studio, 1540 Keller Pkwy #108-261, Keller, TX 76248 or nearnormaldesign@gmail.com.

Using Penny Ante Science Activities

Many schools require that teachers link their instruction to some sort of learning standards. To do this, the teacher will have to provide a connection to the specific standards used by the school. The example below is based on the national standards written by practitioners from around the United States under the auspices of the National Academy of Sciences (NAS) and the National Academy of Engineering (NAE). The *Next Generation Science Standards* (2013) are made up of three Dimensions: Practices, Crosscutting Concepts, and Disciplinary Core Ideas. A Performance Expectation is the point at which these three Dimensions intersect. To access the standards go to: http://www.nap.edu/read/18290/chapter/1. You may purchase the book, download the standards or utilize it for free online. If you use an electronic version, simply choose the topic of the activity that you will teach and put that into the search function. You'll be given all of the entries for that topic; you must decide upon the objective that best fits your lesson.

Depending on how the instructor decides to use the activities, in particular the follow-up discussion of the answers to the questions associated with the activities, student learning could fall into the categories listed on this table. However, every activity could be a part of another Disciplinary Core Idea. For instance, Adaptation and Survival could be used to teach graphing, the history of science, use of camouflage, and predator/prey relationships.

Activity	Performance Expectation	Science and Engineering Practices	Disciplinary Core Ideas	Crosscutting Concepts
Scents or Sense (critical thinking)	2-PS1-1 Plan and conduct an investigation to describe and classify different kinds of materials by their observable properties.	2-PS1-2 Analyzing and interpreting data	PS1.A Structure and properties of matter	2-PS1-1 Patterns in the natural and human designed world can be observed.
Great Buffalo Shortage (earth/space science)	MS-ESS3-4 Construct an argument supported by evidence for how increases in human population and per-capita consumption of natural resources impact Earth's systems.	MS-ESS3-4 Engaging in arguments from evidence	ESS3.C Human impacts on Earth systems	MS-ESS3-1, MS-ESS3-4 Cause and effect relationships may be used to predict phenomena in natural or designed systems.

©2016 Near-Normal Design and Production Studio

Adaptation and Survival (life science)	HS-LS4-4 Construct and explanation based on evidence for how natural selection leads to adaptation of populations.	HS-LS4-2, HS-LS4-4 Constructing explanations and designing solutions	LS4.C Adaptation	HS-LS4-2, HS-LS4-4, HS-LS4-5, HS-LS4-6 Empirical evidence is required to differentiate between cause and correlation and make claims about specific causes and effects.
Hot Tamales (physical science)	MS-PS1-2 Analyze and interpret data on the properties of substances before and after the substances interact to determine if a chemical reaction has occurred.	MS-PS1-2 Analyze and interpret data to determine similarities and differences in findings.	PS1.B Chemical reactions	MS-PS1-4 Cause and effect relationships may be used to predict phenomena in natural or designed systems.

Table of Contents

Using Penny Ante Science Activities 4

Problem Solving
Buttons, Buttons, Who's Got the Buttons?? 11
The Button Bingo Challenge 14
Paper, Paper . . . Which Is the Best Buy? 17
Scents or Sense 19
Sticks and Stones 21
Tools of the Trade 22
What Am I? 25
Which Way Did They Go? 26

Earth and Space Sciences
Baguios and Willy-Willies 31
Catch Me if You Can 33
Chili Today, Hot Tamale 34
Closed Cars, Plants, and Venus 36
Energy Consumers 38
Fault? Whose Fault? 45
Go Fly a Kite 52
Going Up 55
The Great Buffalo Shortage 57
How High Are the Clouds? 59
Humpty Dumpty and the Chocolate Chip Mountain 61
Insulation and Energy 63
Moonstruck or Moonshine? 65
Oh My! It's Full of Stars 67
Only the Shadow Knows 69
Radioactive Decay: A Dating Game! 71
Simple Tests, Simple Answers 73
Splish, Splash 77
Sunrise, Sunset 79
Swirls and Curls 81
Temperature: The Oceans' Pathfinder 83
Them Bones, Them Bones! 86
Trains, Race Cars, and Sirens 92
Water, Water, Everywhere ... But How Much? 93
Where Do I Get My Energy? 96

Life Sciences
Adaptation and Survival 101
Air Pollution 103
Alive Or ??? 105
The Beat Goes On: Part 1 106
The Beat Goes On: Part 2 107

©2016 Near-Normal Design and Production Studio

Blink or Wink	109
Designer Genes!	111
Feast or Famine?	113
Fuzzy or Furry	115
Hot or Cold?	118
Keep in Touch	121
Make a Place	123
More Designer Genes	125
Name Game	128
Presto changO!	130
The Rag Man	132
Right Eye, Left Eye, Which Eye?	134
Sherlock Who?	136
Smoke Signals	138
Survival of the Fittest!	140
Water Pollution	142
What's in It?	144
Where Did You Get It?	146
Whose Eyes?	151
Wink or Blink	154
Wonder Walk	155

Physical Sciences

Balloons and Rice Krispies®	159
Boondoggle Derby!	160
Bridge over the River "Wide"	162
Deep Water	163
Dense, Who's Dense?	165
Does Gravity Work?	167
Drip, Drip, Drip . . .	169
Drip, Drip, Drip Challenge!	171
Eureka!! I've Found It!	173
Furlongs, Leagues, Stones and Pennyweights!!	174
Go? No Go?	176
The Great American Egg Drop	177
Hot Tamales	179
Leaky Pipes	181
Leaning Tower of Pizza!	182
Magnets and "Attractive" Forces	184
Match Me if You Can!!	186
Mirror Images, segamI rorriM	187
Mirror, Mirror on the Wall	189
Screwdrivers, Big and Small	191
Solid to Liquid, Ice to Water	193
Solutions and Mixtures	194
Sound, Noise, and Bottle Music	196

Standard Weight . . . A Potato!	198
Static and "Electric Doorknobs"	200
Stretch, Bend, Rub or Warp	202
Up, Down, All Around!	204
What Is Full?	205

Buttons, Buttons, Who's Got the Buttons??

Objectives:
    The students will classify objects by comparing similarities and differences.
    The students will describe their groupings.
    The students will give logical reasons for their arrangement of the buttons.

Materials:
    A container of buttons

Procedure and Results:
1. Look at your buttons. How are they alike?

2. How are they different?

3. Arrange your buttons into groups. What are the names of your groups (round, oval, two-hole, four-hole, white, brown or what)?

4. Think of another way to group your buttons. Describe your new groups.

5. Explain why you arranged your buttons as you did.

6. Do you think your button groups will work for all buttons?

7. Try the button wrapped in the tissue in the bottom of your container. Will your groups work for this button? What would you have to do to get this new button to fit into a group?

©2016 Near-Normal Design and Production Studio

8. How do scientists use grouping systems?

9. What happens if something won't fit into a scientist's grouping system?

10. What other types or kinds of buttons might you find?

Buttons, Buttons, Who's Got the Buttons??

Teacher's Instructions

Objectives:
    The students will classify objects by comparing similarities and differences.
    The students will describe their groupings.
    The students will give logical reasons for their arrangement of the buttons.

Materials:
    A container of buttons

Make sure each container has an odd button, such as a one-hole or a square button, wrapped in tissue. Secure the tissue so that it won't slide out with the buttons. As an alternative, you may want to pass out these different "buttons" as students reach step 7.

The Button Bingo Challenge

Objectives:
   Students will develop a classification system.
   The students will describe their systems.
   The students will give logical reasons for their arrangement of the buttons.

Materials: A container of at least 10 buttons

Procedure and Results:
1. Design a classification system for the buttons. You may use attributes such as color, shape, texture, and so forth.
2. What attributes or characteristics did you use?

3. What other divisions might you have used?

4. Will your system classify all types of buttons? How do you know?

5. What about the button hidden in the tissue at the bottom of your container? Describe how it fits into your system. If it won't fit into your system, design another one and describe it.

6. What does this tell us about the term BUTTON?

7. What are some other terms that, yet spelled the same, have different meanings or uses?

©2016 Near-Normal Design and Production Studio

8. What does this tell you about classification systems?

9. What are some changes made by scientists in our classification systems (such as animal and plant classifications or the chemical periodic chart)?

10. Are these classification systems "static" or "dynamic"? Why?

The Button Bingo Challenge

Teacher's Instructions

Objectives:
    Students will develop a classification system.
    The students will describe their systems.
    The students will give logical reasons for their arrangement of the buttons.

Materials:
    A container of at least 10 buttons plus one "button" wrapped in tissue

Make sure each container has a slogan "button", such as "I like Ike!", wrapped in tissue. Secure the tissue so that it won't slide out with the buttons. As an alternative, you may want to pass out these different "buttons" as students reach step 5.

Paper, Paper . . . Which Is the Best Buy?

Objectives:
 Students will design an experiment to determine which toilet paper is the best bargain.
 Students will determine which variables are important for finding the best bargain.

Materials:
 Six or more brands of toilet paper (including pricing information, sheets per roll, and length per roll), eyedroppers, paper towels, water, BBs, a small beaker, and tweezers.

Procedure and Results:
 A recurring dilemma facing anyone trying to be a wise consumer is the determination of which of several brands is the best buy. It is not always the least expensive or the largest size, often other qualities about the brands are important factors in finding out which is the best bargain. These variables require investigation.
 In the case of toilet paper, the variables could be things like sheets per roll, single or double sheets, rough or smooth texture, square feet per roll, strength, and the like.
 In any scientific research, the investigators must control as many variables as possible. Today, your task is to devise a series of tests to determine which among your brands of toilet paper is the best value and why.

1. Criteria for study should include:
   - Cost (measured in cents per sheet, cents per square foot, or some other way)
   - Strength (relative to other brands) You may want to test them wet and dry.
   - Absorption (relative to other brands)

2. Work in your group and decide how you will tackle each variable. Keep a record of the tests you do and how you did them. Be prepared to explain and defend your methods.

3. Complete the chart with the rest of the class.

	Brand A	Brand B	Brand C	Brand D	Brand E	Brand F
Cost						
Strength						
Absorption						

©2016 Near-Normal Design and Production Studio

4. Now give each brand a ranking (1 = best, 6 = worst) for each of the tests.

	Brand A	Brand B	Brand C	Brand D	Brand E	Brand F
Cost						
Strength						
Absorption						

5. The brand with the lowest total ranking is the best buy. That is, a brand that came in first among all five tests would have a total of 3; one coming in last in each test would have a score of 18. Which of your brands is the winner?

6. Which variable(s) were most important in finding out which toilet paper is the best bargain? Why?

7. What are other variables you should consider when you purchase toilet paper? Why are these important?

Scents or Sense

Objectives:
　　Students will classify objects based on their characteristics.
　　Students will describe objects using their sense of touch.

Materials:
　　Eraser, nail, carpet sample, sandpaper, sponge, lead weight

Procedure and Results:
1. Close your eyes as your partner places objects in front of you.
2. With your eyes closed, feel of each object.
3. Pick up the object that is hard. Which object is it?

4. Pick up the object that is soft. What object is it?

5. Pick up the object that is light. What object is it?

6. Pick up the object that is heavy. What object is it?

7. Pick up the object that is smooth. What object is it?

8. Pick up the object that is rough. What object is it?

9. Trade roles with your partner and repeat steps 1-4.
10. What do we mean by the terms hard, light, and smooth?

©2016 Near-Normal Design and Production Studio

11. Classify the objects according to the attributes of hardness, lightness, and smoothness. Which are hard?

12. Which are light?

13. Which are smooth?

14. Are there any objects that fit into more than one of these groups? If so, what are they and into which groups do they fit?

15. What other attributes could you use to classify these objects?

16. Why is important to be able to classify things by touch?

Sticks and Stones

Objectives:
   Students will classify objects by comparing similarities and differences.
   Students will apply classifications to environmental processes.

Materials:
   Samples of coal, leaves, sugar, animal, salt, calcium, sand, cement, and Center Area

Teacher's Procedure and Results to Discuss:
1. Discuss with students the words organic and inorganic. Explain that the term organic refers to objects that are either alive or were once alive. Also explain that inorganic describes things that never lived.

2. Ask them to identify objects in the room that they think are organic and inorganic.

3. Allow the students to come to the Center Area and organize the materials into the categories "organic" and "inorganic."

4. Ask the students why they organized the objects in the manner selected.

Tools of the Trade

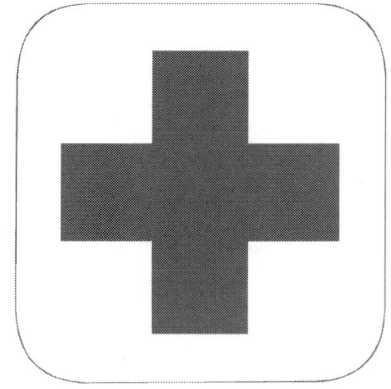

Objectives:
    Students will identify laboratory safety materials and equipment.
    Students will determine which signs apply to practicing science.

Materials:
    Forceps, latex gloves, safety goggles, hot mitts, safety symbols (road signs and laboratory signs)

Teacher's Procedure and Results to Discuss:
1. Explain to the students the roles of signs in our world.
2. Show by example how scientists use the various instruments and materials.
3. Using the road signs, (i.e. Stop, Yield, road direction) ask the following questions:
   What do these signs mean?

   Why do we use these signs?

   Do the shapes and colors have special meaning?

4. Show the science signs.
   What do you think words such as "ACID", "BASE", "RADIATION" and "POISON" mean?

   Do the colors and shapes have special meaning?

   Why do we need to know what these signs mean?

5. Which signs do you think are most important when you are doing science? Why?

Tools of the Trade Signs

Tools of the Trade Signs

ACID

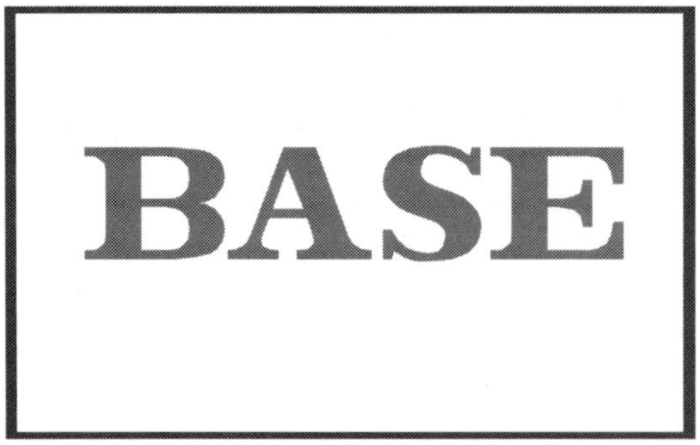

BASE

What Am I?

Objectives:
   Students will describe shapes.
   Students will match shapes to one another.
   Students will group shapes by more than one characteristic.

Materials:
   A variety of large, colored shapes (one set for the teacher and at least two matching sets to share among the class)

Teacher's Procedure and Results to Discuss:
1. Give each child a shape.
   Hold up a shape and have the children with the same shape stand up.
   Ask the children to trace the shape with their finger as they repeat its name.

2. Play a game.
   Give each child a set of shapes and have him/her match the shapes with ones you have placed on the floor.
   The difficulty of this activity can be increased by asking them to match shapes and color, shapes and size, shape and texture (if available), color and size, and so forth.

3. Ask the students to name the characteristics of their shapes, such as color, size, shape, texture, and other attributes.

4. Ask students to name other objects that have some of these shapes.

Which Way Did They Go?

Objectives:
    Students will give and follow directions.
    Students will observe patterns.

Materials:
    2 each - keys, paper clips and pennies

Procedure and Results:
1. Give your lab partner one of each of the keys, pennies and paper clips.
2. Hide your key, penny and paper clip from your partner's view. Arrange them in some pattern.
3. With your partner unable to see your pattern, give him/her directions as to how to arrange the key, penny and paper clip so as to duplicate your pattern.
4. How did it go?

5. What are some observations your **partner** had to make?

6. Now have it's your partner's turn to take the lead. Let him/her hide the key, penny and paper clip and arrange a pattern. This time he/she will not give directions but will answer questions with a "yes" or "no" response as you try to duplicate his/her pattern.
7. How did it go this time?

8. What are some observations **you** had to make?

9. Once again, you arrange a pattern. Be careful that your partner cannot see your pattern. This time as your partner asks questions, respond with "non-verbal" responses, such as nodding of the head, and so forth.
10. How did it go this time?

11. How does this relate to the way we give and receive directions from each other?

12. Which method did you find caused the greatest problem?

13. How does this relate to the day-to-day work of being a scientist?

14. How does this relate to the daily work of being a forest ranger?

Earth – Space Sciences

Baguios and Willy-Willies

Objectives:
Students will determine the movement of air in a cyclone and an anticlone.
Students will explain the difference between low and high air pressure.
Students will relate air pressure to weather systems.

Materials:
For every pair of students: small aluminum pie pan, 35cm x 35cm cloth, stapler, scissors, split fishing weights, metric ruler

Procedure and Results:
1. Cut the bottom from the pie pan, leaving a 0.5cm edge.
2. Fit the cloth into the pan as you would a piecrust, and staple it to the sides of the pan, near the bottom.
3. Put the weights in the center of the pan and push up from underneath gently. Where do the weights go?

4. What type of pressure area does this represent?

5. Where does the air get pushed to in a low pressure area?

6. Gently lower the cloth. Where do the weights go?

7. What type of pressure does this represent?

©2016 Near-Normal Design and Production Studio

8. Where does the air get pushed to in a high pressure area?

9. Which movement of weights represents a cyclone? How do you know?

10. Which movement of weights represents an anticlone? How do you know?

11. In what part of the world would you find anticlones?

12. What are Baguios and Willy-Willies?

Catch Me if You Can

Objectives:
   Student will identify objects attracted to a magnet.
   Students will relate magnetism to certain earth materials.

Materials:
   Small magnets, variety of objects such as paper clips, eraser, rubber band, penny, magnetite, other non-magnetic rock, cups (paper or Styrofoam), labeled "YES" and "NO," a learning station area

Teacher's Procedure and Results to Discuss:
1. Give each child a magnet.

2. Have them take their magnet to the learning station and experiment with the objects, placing those attracted to the magnet in the "YES" cup and those not attracted in the "NO" cup.

3. Ask children to tell how the things that were attracted to the magnet are alike. How are these different from those that were not attracted to the magnet?

4. Talk about why some things are attracted and others not. Be sure to reinforce that only certain metals are attracted to magnets and that these metals come from rocks within the earth.

5. Ask the students to tell how magnets are useful to man and where they might use them at home.

Chili Today, Hot Tamale

Objectives:
  Students will measure temperature using a thermometer.
  Students will hypothesize as to why temperatures of various surfaces differ.
  Students will determine the relationship between temperature and heat.
  Students will relate temperature to climate.

Materials:
  Thermometers

Procedure:
1. Choose at least 5 teams of students.
2. Read thermometer and record temperature on the data table before going outside.
3. Place the thermometers in the following areas (make sure these areas are out of normal traffic paths so that the thermometers do not get broken):
   Team 1--in the sun on the sidewalk
   Team 2--in the sun on the blacktop playground or parking lot
   Team 3--in the sun on a grassless area
   Team 4--in the sun on the grass
   Team 5--in the shade on the grass.
4. After 5 minutes, read the thermometer and note the temperatures on the data table. Copy the data from the other teams to your data sheet.

Results:

	Inside Temperature	Surface Temperature	Difference
Team 1			
Team 2			
Team 3			
Team 4			
Team 5			

1. Were all the beginning temperatures the same?

2. Why were some of them different?

3. Were all the ending temperatures the same?

4. Which was the hottest area?

5. Why do you think this is?

6. Which was the coolest surface?

7. Why do you think this is?

8. If you designed a building, what materials would you want around it to keep it cool?

9. What if you lived in a colder climate? How would you design the area around a building?

10. What is the difference between heat and temperature? How do you know?

Closed Cars, Plants, and Venus

Objectives:
   Students will read a thermometer.
   Students will operationally define the greenhouse effect.
   Students will relate the greenhouse effect to changes in weather.
   Students will examine the relationship between the abiotic and biotic portions of the environment.

Materials:
   For each pair of students: 2 thermometers, 2 large foam cups, 1 plastic bag, and 1 rubber band.

Procedure:
1. Cover one of the cups with the plastic bag. Use the rubber band to hold the bag in place.
2. Put a thermometer in each cup (you'll have to poke a hole in the bag).
3. Read the thermometers and record the temperatures on the chart below. Set the cups in the sun.
4. Read the temperatures again after 10 minutes and after 20 minutes. Be sure to record your data.

Data:

Time	Open Cup Temperature	Closed Cup Temperature
0 minutes		
10 minutes		
20 minutes		

Results:
1. Which cup got the hottest?

2. In which cup did the temperature rise first?

3. Which cup is like a greenhouse? Why?

©2016 Near-Normal Design and Production Studio

4. Describe the greenhouse effect.

5. Why is the greenhouse effect important to earthlings?

6. Why is it not a good idea to leave a pet in the car, especially in the summer?

Energy Consumers

Objectives:
  Students will interpret the pie chart.
  Students will form hypotheses regarding energy usage.
  Students will relate ethics to energy usage.
  Students will determine the impact energy usage has on the environment.

Materials:
  100 fish shaped crackers, 100 smiley faces

Procedure:
1. Using the World Energy Usage table to construct a pie chart, then determine how many fish shaped crackers each geographic area will get, for example North America gets 25 crackers.
2. Distribute the crackers based on the findings in step 1.
3. Who gets the most crackers?

4. Who gets the fewest?

5. If technological advancement is equivalent to energy consumption, which area is the most technologically advanced?

6. Rank order them from 1 = most advanced to 8 = least advanced.
    1. _____
    2. _____
    3. _____
    4. _____
    5. _____
    6. _____
    7. _____
    8. _____

7. What conclusions regarding technological advancement can you draw related to energy consumption?

8. Suppose you had 1000 fish shaped crackers instead of only 100. How many crackers would Western Europe receive?

9. How many would the geographic region of your origin receive?

10. If you divided the fish by population instead of by energy usage, what do you think the rank order of the countries would be (1 = most populated to 8 = least populated)? Make a guess.
    1. _____
    2. _____
    3. _____
    4. _____
    5. _____
    6. _____
    7. _____
    8. _____

11. Use the World Population table to construct a pie chart. Your teacher will distribute the "Smiley" faces based on population to each of the regions. Compare your numbers of "Smiley" faces with the amount of energy you have to distribute using the pie charts you made.

12. How does your geographic region's population compare to its energy use?

13. How does your region's population compare to its technological advancement?

14. What general statements can you make about the distribution of energy, locations of population densities, and technological advances?

15. How do you think these distributions interact with other societal questions such as civil rights, education, hunger, and disease?

16. Find the source of each region's energy.
    1. _____
    2. _____
    3. _____
    4. _____
    5. _____
    6. _____
    7. _____
    8. _____

17. How do these sources compare with the region's technological abilities?

World Energy Usage

Location	Percent
Africa	3%
Asia	22%
Europe	22%
Middle East	5%
North America	25%
Pacifica	2%
Russia and the Former Soviet Union	16%
South and Latin America	5%

Updated to reflect 2016 levels

World Population

Location	Percent
Africa	22%
Asia	39%
Europe	10%
Middle East	4%
North America	8%
Pacifica	3%
Russia and Former Soviet Union	5%
South and Latin America	9%

Updated to reflect 2016 levels

Energy Consumers
Teacher's Instructions

Objectives:
    Students will interpret the pie chart.
    Students will form hypotheses regarding energy usage.
    Students will relate ethics to energy usage.
    Students will determine the impact energy usage has on the environment.

Materials:
    100 fish shaped crackers, 100 smiley faces

Procedure:
1. Count off the students into nine groups.
2. Randomly assign the name of each geographic region to a group.
3. You may want to prepare two transparencies with a pie chart on each.
4. You may also want to have the bags of goldfish and groups of "Smiley" faces prepared rather than have the students count out their own.

Fault? Whose Fault?

Objectives:
   Students will observe a model of natural phenomena.
   Students will make and use a model to form hypotheses.
   Students will determine the forces necessary to produce changes in rock layers.
   Students will describe the structure of the rock layers.

Materials:
   Blank wooden fault blocks (2" X 4" with two 30° cuts, sanded), paint pens, pattern or tag board pattern, permanent markers, scissors, tape

Procedure for wooden blocks:
1. Examine the sample blocks. Decide how you will design yours and proceed (pencil first).
2. When you are satisfied, complete and color your blocks.

Procedure for tag board blocks:
1. Look closely at the pattern, noting that each piece has the same design but two are in different order.
2. Color each design with the same shade.
3. Cut out the patterns and fold them along the lines. Tape the sides together. You will have three blocks.

Results:
1. What do the different colors represent?

2. With your blocks in the following arrangement, practice what happens when you exert the indicated forces:

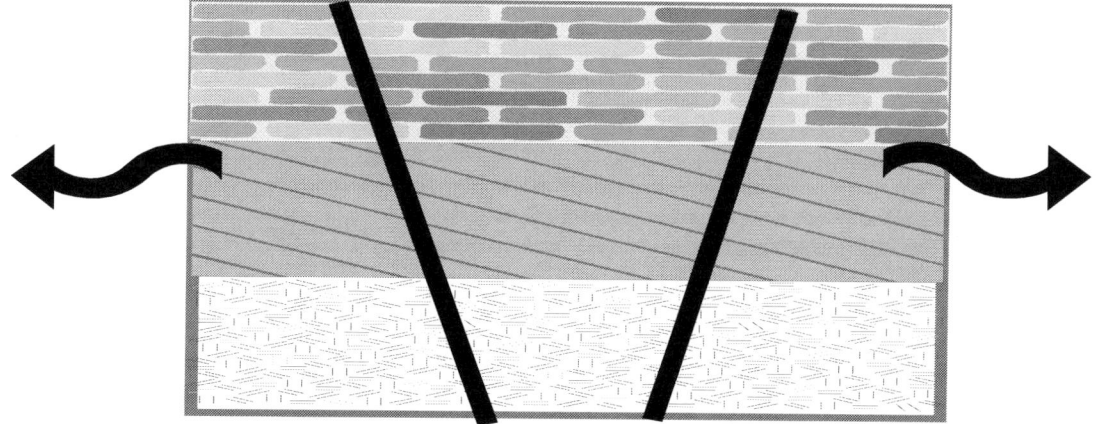

3. What happens? Describe and draw the results.

4. When extension occurs, part of the rock layers fall down resulting in a normal fault. The footwall (an area you could walk up) is above the fallen block. Have you created a normal fault? How do you know?

5. Arrange your blocks as shown and exert the forces:

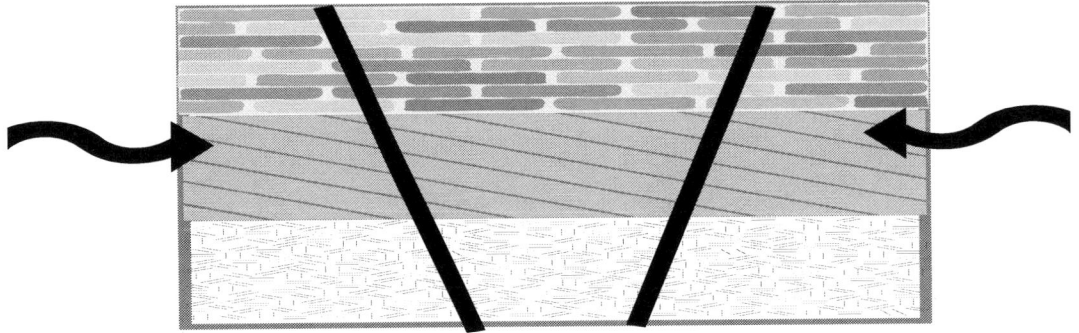

6. What happens when the indicated forces are exerted?

7. Illustrate your observations:

Fault? Whose Fault?

8. When compression occurs, part of the rock layers push up resulting in a reverse fault. The footwall (an area you could walk up) is below the lifted block. Have you created a reverse fault? How do you know?

9. Using only the two end blocks, produce a normal and a reverse fault. Draw and label your models.

10. In both a reverse and a normal fault, hanging walls are also produced. The hanging wall in a normal fault is the wall at the bottom of the fallen block. The hanging wall in a reverse fault is on the lifted block and hangs above the other block. Label the hanging walls in your above drawings.

11. Would you want to build a house near a fault line? Why?

12. Explain what happens to the rock layers when faulting occurs.

13. How could faulting affect agriculture?

14. How is the model you made like a real fault? How is it different from a real fault?

©2016 Near-Normal Design and Production Studio

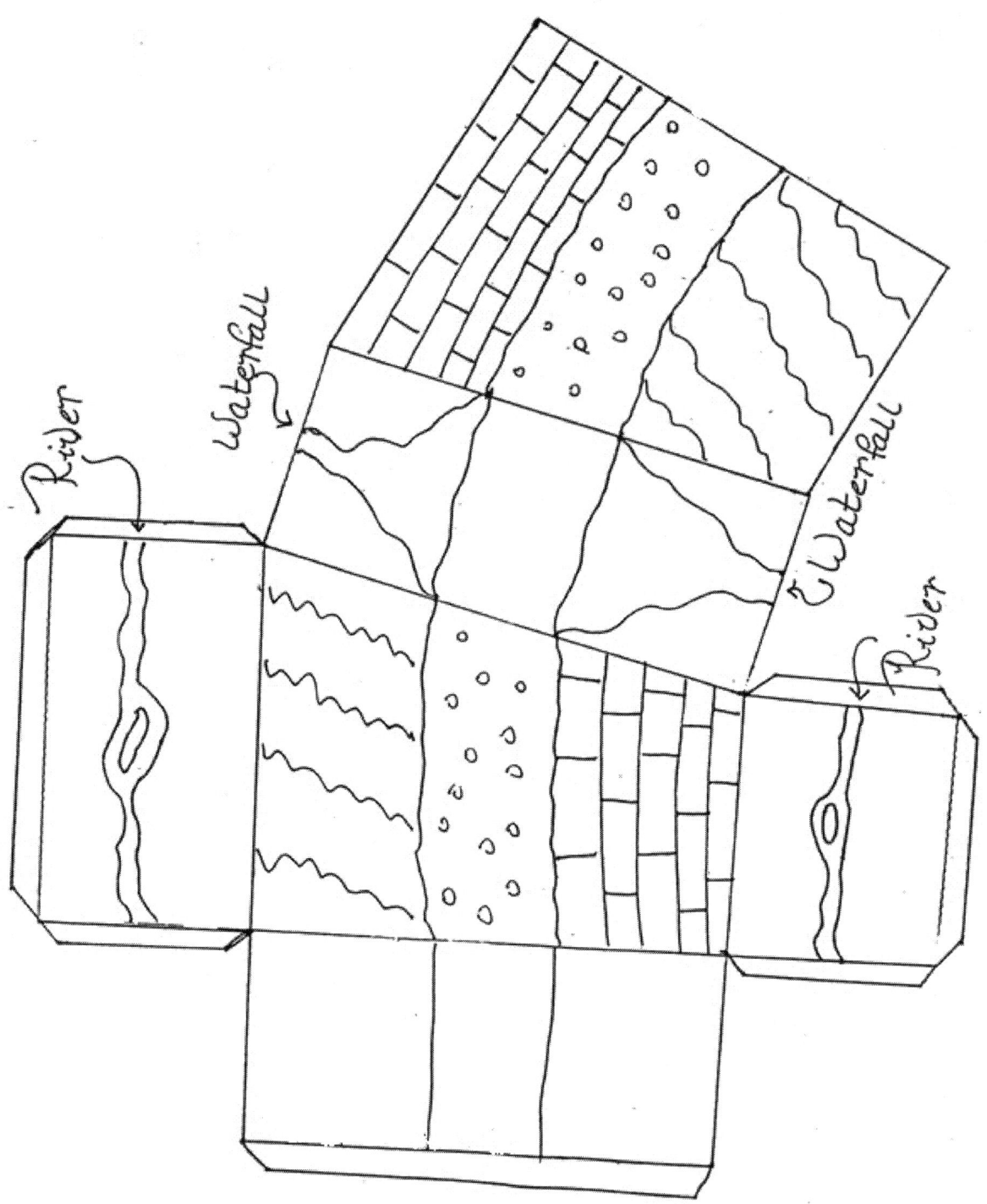

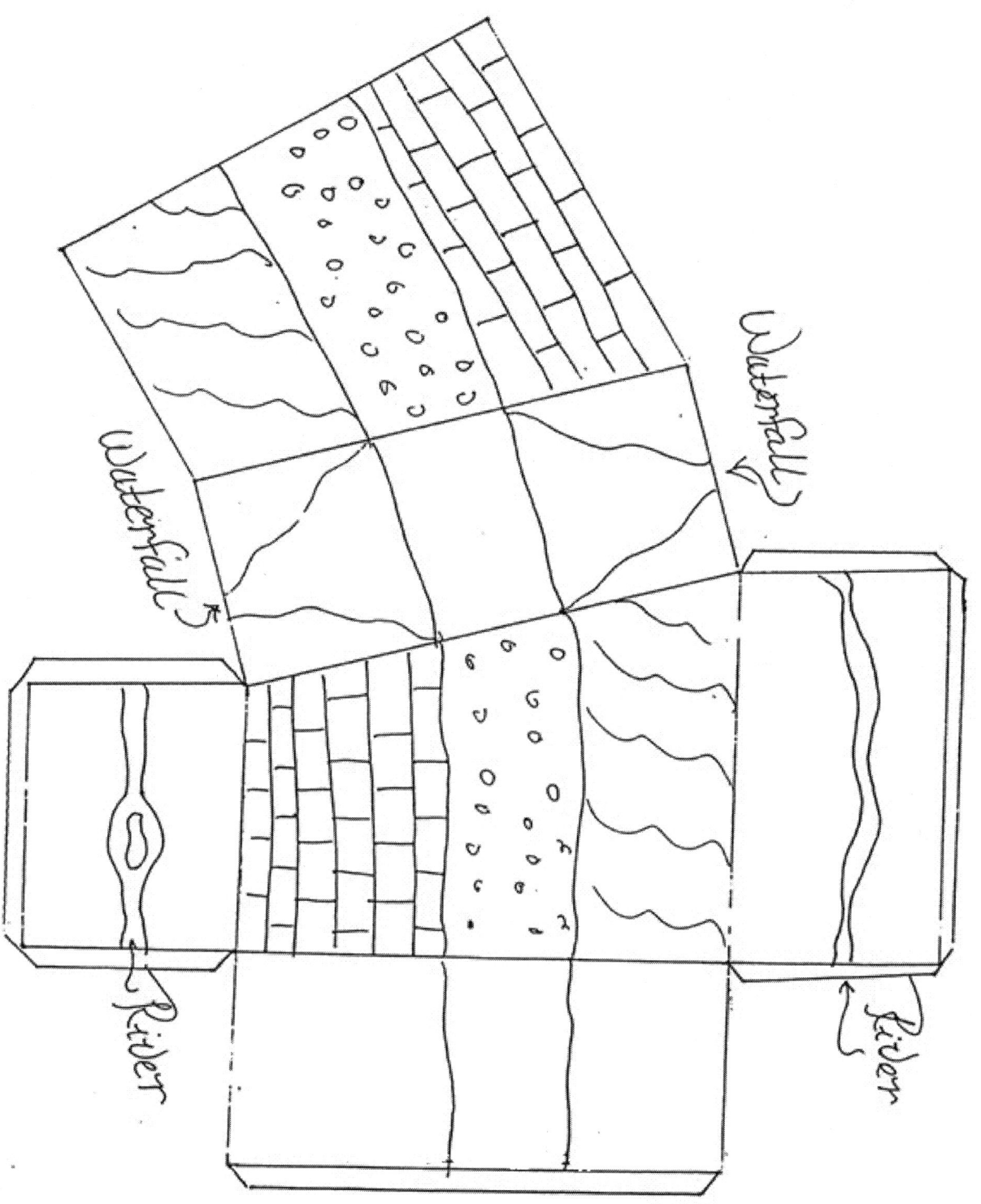

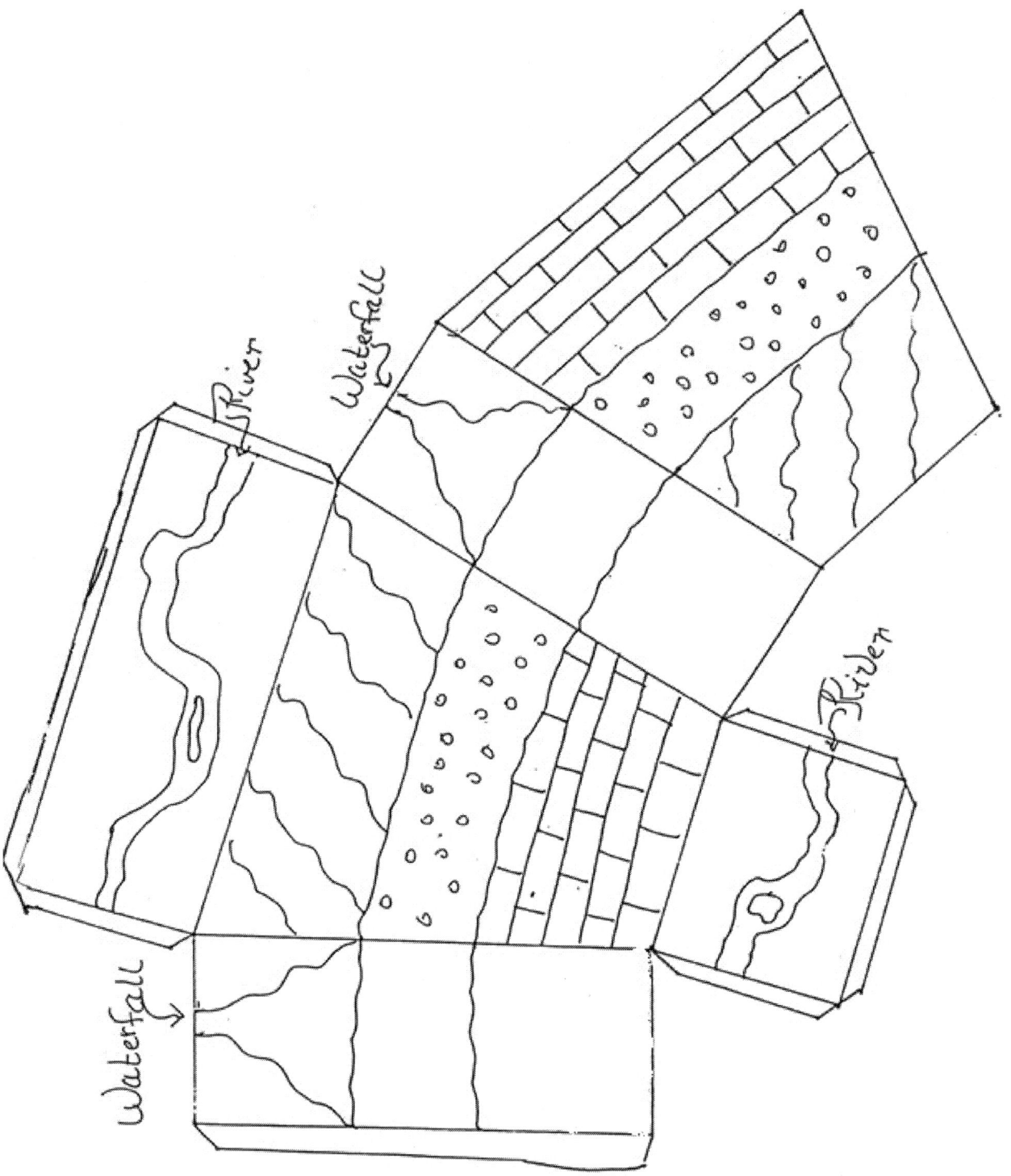

Fault? Whose Fault?
Teacher's Instructions

Objectives:
    Students will observe a model of natural phenomena.
    Students will make and use a model to form hypotheses.
    Students will determine the forces necessary to produce changes in rock layers.
    Students will describe the structure of the rock layers.

Materials:
   Blank wooden fault blocks (2" X 4" with two 30° cuts, sanded), paint pens, pattern or tag board pattern, permanent markers, scissors, tape

Procedure for wooden blocks:
1. Prepare the blocks by sanding them smooth.
2. Use the paper pattern as a guide to color the wood. DO NOT USE WATERBASED MARKERS.
3. The front of the model is different from the back. Turn one of the end blocks over and fit it to the front of the other end block (the angles of the blocks should match so that when they are put together correctly, it looks like a solid piece of wood). Copy the pattern onto the blank block and color it.
4. Turn the other two blocks over and match them to the block you have just colored. Follow the pattern on these other two pieces of wood.
5. Paint rivers on the top and bottom of the blocks. Add waterfalls on the edges of the blocks.
6. Before students paint the backs of their models, you may want to check their designs to make sure layers will match.

Procedure for the tag board blocks:
1. Copy the patterns onto tag board. Note that there are three different patterns.
2. Shade matching layers the same color.
3. Cut out the models, fold along the solid lines and tape them together.

©2016 Near-Normal Design and Production Studio

Go Fly a Kite

Objectives:
  Students will define renewable and non-renewable resources.
  Students will predict how wind will affect flight.

Materials:
  24-soda straws per group, string, tissue paper (and pattern), glue stick, scissors

Procedure:
1. Construct four tetrahedrals according to the following instructions.

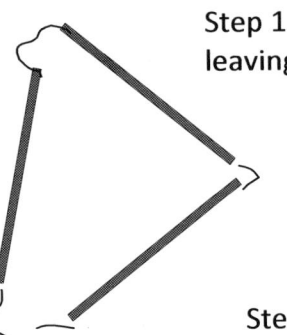

Step 1: Pass the string through three straws leaving a 4cm tail.

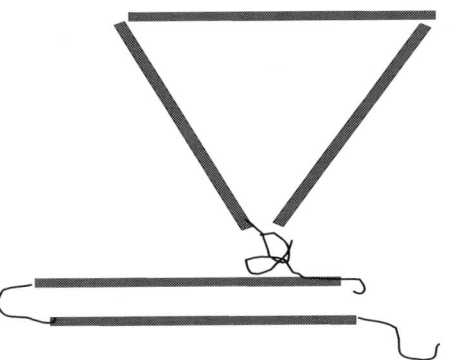

Step 2: Tie a tight knot to form a triangle.
Step 3: Draw the long tail of the string through 2 more straws.

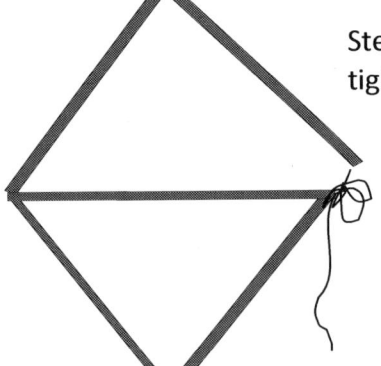

Step 4: Form another triangle and tie the string tightly.

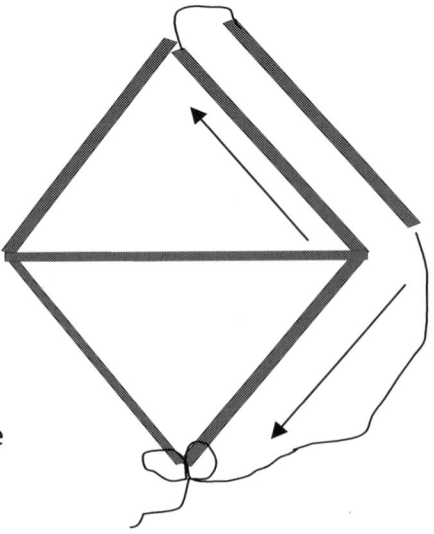

Step 5: Send the string back up one of the sides.
Step 6: Add another straw and tie it tightly to the point of the other triangle. You have a tetrahedral.

©2016 Near-Normal Design and Production Studio

2. After constructing four structures, cover 2 sides of each with tissue paper using glue sticks.
3. Now you can assemble all four structures into a single structure. Several of these new structures can be assembled together to form even larger kite structures.

Results:
1. How does the wind lift the kite? Draw the kite and show how the wind will lift it. You may use arrows to represent the wind.

2. How might have the wind been harnessed at first for use by man?

3. Why do we consider the wind to be a renewable energy resource?

4. List other renewable resources.

5. What are some nonrenewable sources of energy? Why are these considered nonrenewable?

6. What might you, as a citizen of this planet, do to encourage use of renewable energy sources?

Go Fly a Kite
Teachers Instructions

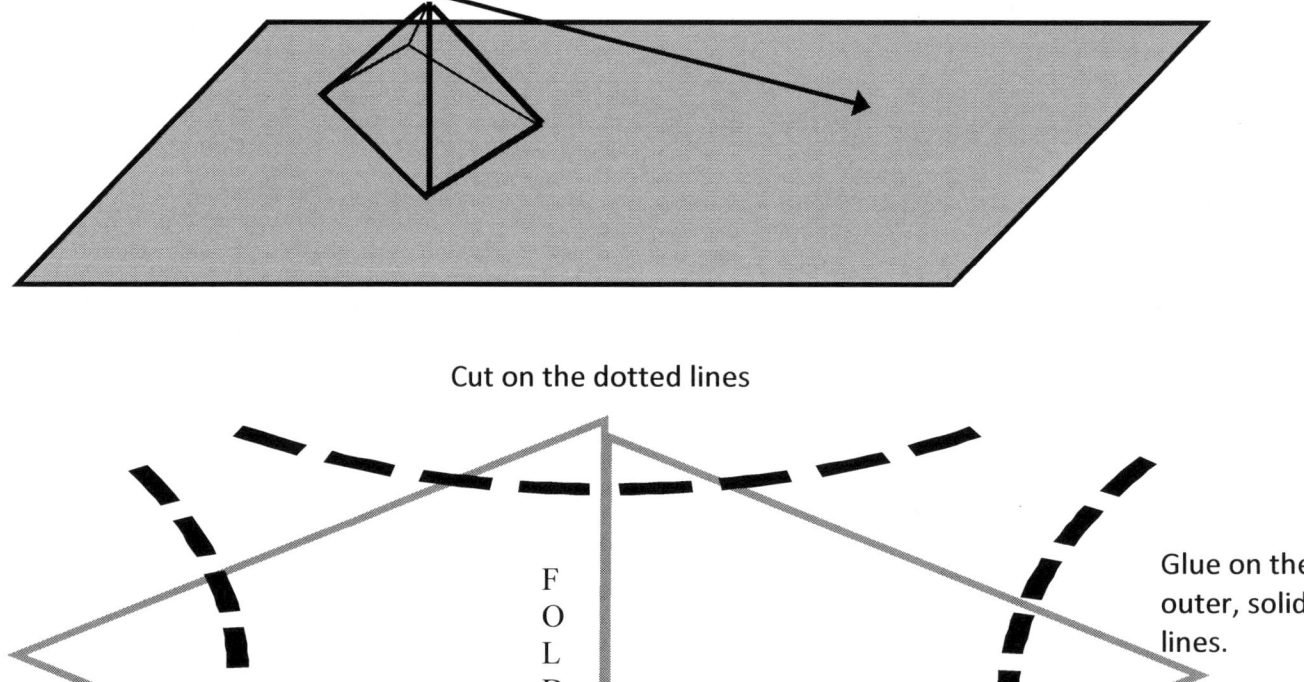

Once students have constructed the tetrahedrals, two sides will be covered with one sheet of tissue paper. An example of the pattern is drawn below. To get the actual size, set the tetrahedral on a sheet of legal size paper, draw an outline (about a half inch larger) around one side. Then, without lifting the tetrahedral, turn it directly over and draw in the same manner around the other side. Cut rounded notches at each corner. Put glue on the half-inch border.

When tying the tetrahedrals together (after the tissue paper has dried), make sure that they are all facing the same direction, otherwise they won't catch the wind correctly.

Cut on the dotted lines

F O L D

Glue on the outer, solid lines.

Going Up

Objectives:
  Students will observe the movement of air currents.
  Students will describe the interactions of air currents and weather.

Materials:
  Commercial bubble liquid, bubble wand, pencil, paper, and a timer

Procedure:
1. Blow bubbles. Note the path the bubbles follow through the air.
2. Write down the direction that the bubbles follow about every ten seconds. Try to estimate their distance from you at the end of 50 seconds. Repeat the procedure five times.

Results:

	10 sec.	20 sec.	30 sec.	40 sec.	50 sec.	distance
Trail 1						
Trial 2						
Trial 3						
Trial 4						
Trial 5						

1. Which direction did the bubbles go at first?

2. As the bubbles got higher, did they change directions?

3. When you stopped timing, how far were the bubbles from you?

4. When did the bubbles pop?

5. Were the bubbles low or high when they popped?

6. Were the bubbles affected by buildings or trees? If so, how?

7. Were the bubbles affected by the type of ground (concrete, grass, etc.) in the area? If so, how?

8. What can you say about wind currents in your area?

9. How are wind and weather related? How do you know?

The Great Buffalo Shortage

Objectives:
  Students will define renewable and non-renewable resources.
  Students will predict results of over use of non-renewable resources.
  Students will graph the expenditure of resources as a function of users.
  Students will determine the impact of energy usage on the environment.

Materials:
  Large box (at least 50cm X 30cm), Styrofoam packing materials to fill the box, 20 household sponges cut into 2cm cubes

Procedure:
1. The box filled with Styrofoam represents the earth. The sponges represent the ore. Each miner has 10 seconds to get as much ore as possible using his/her non-dominant hand. The 10 seconds represents 1 year. It takes 10 nuggets/year to make a profit.
2. In the first year there is one miner. Be sure to record the number of nuggets mined on your graph. **Do not return the nuggets to the earth.**
3. Miners who are careless about ecology (as indicated by the mess on the floor) will be assessed a damage tax of 20% of their total production for that year.
4. The second year there are two miners. Graph their total production.
5. There are four miners the third year. Be sure to graph total production for each year.
6. The fourth year there are 8 miners.
7. The fifth year there are 16 miners.
8. Continue doubling the number of miners until the total production rate drops to about what it was the first year.

Results:

1. What was the largest number of nuggets per person mined?

©2016 Near-Normal Design and Production Studio

2. What was the smallest number of nuggets per person mined?

3. What is the relationship between number of miners and available resources?

4. What type of resource were the nuggets, renewable or non-renewable? How do you know?

5. If each year the nuggets had been replaced in the earth, what type of resource would they be? How do you know?

6. If oil is like the nuggets, what will happen if we all use as much as we want?

7. If there had been no environmental damage tax, what would have happened to the environment?

8. What is a renewable resource? List 4 renewable resources.

9. List 4 nonrenewable resources.

How High Are the Clouds?

Objectives:
  The student will measure the height of the clouds.
  The student will read a thermometer accurately.
  The student will determine the relationship between clouds and dew point.

Materials:
  For every pair of students: empty metal can (smoothed where the lid has been removed), 2 thermometers, 100ml ice, 100ml water, 50cc salt, rubber band, waste bucket

Procedure:
1. Go outside to an open area.
2. Attach the one of the thermometers to the outside of the can using the rubber band.
3. Use the other to take the temperature of the outside air. Be sure to enter this on your data table.
4. Put the ice, salt, and water in the peanut can. Observe the can closely.
5. When moisture forms on the surface of the can, read the temperature of the can. Record this on your data table.
6. Dump the ice, salt and water into the waste bucket. Observe the can carefully.
7. When the moisture disappears, read the temperature again, and record the data.

Results:

Air Temperature	
1st Reading with Ice	
2nd Reading after Ice	

1. What is the moisture on the outside of the can called?

2. What is the moisture that forms on the grass in the morning called? How is the moisture on the can like that on the grass?

3. Average the first and second temperature readings, then use this number in the formula below.

$$\text{Cloud height} = \frac{\text{Air temperature - Average}}{5.5} \times 1000 \text{ feet}$$

Cloud height = _____

4. How do you know when the air reaches the dew point?

5. What has the dew point to do with the height of clouds?

Humpty Dumpty and the Chocolate Chip Mountain

Objectives:
    Students will identify factors associated with man's effect on the environment.
    Students will make a model of a mine.
    Students will state the differences between renewable and non-renewable resources.

Materials:
    Chocolate chip cookies (two per student), napkins or paper towels, paper clip

Procedure:
1. Give each student two cookies (one to eat and one to experiment with). They will also need a napkin and/or paper towel and paper clip.
2. Place the cookies on the napkin, top up; **you may not turn your cookie over**. With your "excavation tool" (the paper clip), start mining the visible "resources" (the chocolate chips).

Results:
1. Count the whole chips you got out of your "mine". Six average size fragments count as one chip. How many total chips did you get out? Report this data to the class by giving the information to your teacher.

2. When you have completed the mining operation, examine the remaining "land mass" in terms of what can or should be done to reclaim the "land". What could you do?

3. In what ways might this be similar to the reclamation plans used by strip mining companies today?

4. Did your "strip mining operation" locate any hidden resources? How might they be reached?

5. Who "mined" most of the resources?

6. What remained of the "Chocolate Chip Mountain" after the mining?

7. What are some environmental issues raised by strip mining?

8. If this cookie had been a chocolate chip, raisin, oatmeal combo, what would determine if this were a chocolate chip mine, a raisin mine, or an oatmeal mine?

9. How is your cookie like a real mine? How is it different from a real mine?

10. Were the resources you removed renewable or nonrenewable? How do you know?

©2016 Near-Normal Design and Production Studio

Insulation and Energy

Objectives:
   Students will construct a temperature versus time graph reflecting the data.
   Students will read a thermometer.
   Students will determine how insulation affects energy usage.
   Students will compare insulation materials.
   Students will state how the use of insulation affects the environment.

Materials:
   For each group of four students: 4 identical soft drink cans, 4 thermometers, clock with second hand, hot water, funnel, newspaper, aluminum foil, 2 foam cups large enough to hold the drink can, masking tape, graph paper, colored pencils.

Procedure:
1. Wrap one can with a thick layer of newspaper. Be sure to leave a hole in the top so that the water can be poured in and the thermometer inserted. Tape the newspaper so that it won't come loose.
2. Wrap a second can with the foil, shiny side out.
3. Place one can inside a foam cup. Poke a hole in the bottom of the other cup. Use this cup for a lid.
4. Leave the fourth can uncovered.
5. Using the funnel, fill all the cans with hot water. Insert the thermometers and take temperature readings.
6. Record the beginning temperatures and temperatures every 2 minutes for the next 20 minutes. You will use these data to make your graph.

Results:
1. Make a graph of the temperature against time. Use a different color for each can.
2. From your graph, list the cans in order of temperature increase.

3. Does insulation prevent heat loss? How do you know?

4. Which insulator worked the best?

5. Why was no insulation used on one can?

6. How is insulation important to you right now?

7. How does a well-insulated house affect energy usage? How does the energy usage affect the environment?

8. If you were going to construct a container to keep cold in and heat out, would you use any of the materials you tested? Explain your answer.

Moonstruck or Moonshine?

Objectives:
   Students will demonstrate safe use of laboratory equipment.
   Students will classify events according to similarities and differences.
   Students will hypothesize reasons for man's reliance on the phases of the moon.

Materials:
   Softball, flashlight, table top, darkened room

Procedure:
1. By arranging the ball and flashlight as illustrated, you can study the phases of the moon.
2. Label the locations on the table 1-8 as illustrated.
3. Move to each number and draw what you see in the blocks below. Look at the baseball, not directly at the flashlight.

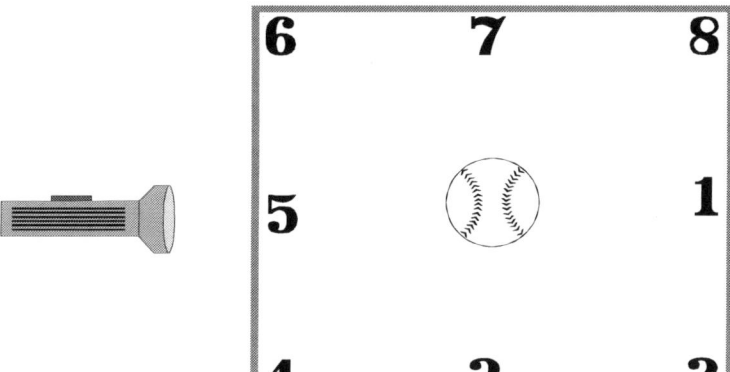

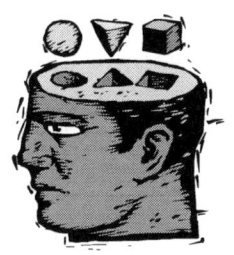

1	2	3	4
5	6	7	8

©2016 Near-Normal Design and Production Studio

Results:
1. Which block was the "full" moon? New moon?

2. Which were the crescents (a bright slice)?

3. Which were the "gibbous" (a dark slice) moons?

4. Which was the first quarter? Third?

5. Where does the moon get its light?

6. Why do we only see part of the moon at any time?

7. Why would farmers be concerned about the phases of the moon?

Oh My! It's Full of Stars

Objectives:
   Students will examine star charts.
   Students will construct a star finder.

Materials:
   Star chart and sky calendar from Astronomy in Your Hands (http://www.astronomyinyourhands.com/starwheel/starwheel.html), a small brad, scissors

Procedure:
1. Follow the directions on the website to make your star finder.

Results:
1. Move the "south" position on the star chart to the bottom of the window. What constellations can you see?

2. Move the star chart to the "west" position. Now which constellations can you see?

3. What other objects can you see?

4. According to the calendar, what planets are visible in the evening sky tonight?

5. Where would these be on your star map?

6. Draw the position of the constellation Cassiopeia relative to the star Polaris. Be sure to label the directions.

©2016 Near-Normal Design and Production Studio

Oh My! It's Full of Stars
Teacher's Instructions

Objectives:
    Students will examine star charts.
    Students will construct a star finder.

Materials:
    Star chart and sky calendar from Cumberland Science
Museum (http://www.skymaps.com/downloads.html), sky window, a small brad, scissors

Procedure:
1. Find the star chart and calendar on the Internet at the address supplied. If this URL is inactive, search under the topic "star chart". There are several other sites available.
2. Copy the star chart and the sky window onto cardstock.
3. You may want to use a hole punch to prepare a hole for the brad.

©2016 Near-Normal Design and Production Studio

Only the Shadow Knows

Objectives:
 Students will determine where the sun rises and sets.
 Students will measure the amount of rotation the earth makes in a given time period.

Materials:
 Sharpened pencil, clothespin, lined paper, protractor, flat sunny patch of ground or window ledge, marker, compass

Procedure and Results:
1. Place the paper on the ground or window ledge and place the sharpened pencil in the clothespin as illustrated. The clothespin is used to make the pencil stand upright. Make sure the shadow of the pencil falls on the paper.
2. Trace this shadow with the marker and label it "A".
3. Do not move the paper. Wait 15 minutes then trace this shadow and label it "B". Wait another 15 minutes; trace and mark this shadow "C".
4. Use the protractor to measure the angle between shadows A and B. The angle is:

5. What is the angle between A and C?

6. The distance around the earth at the equator is about 15625 km. If the earth travels this far in 24 hours, how far will it travel in 15 minutes?

7. How far in 30 minutes?

8. How are the degrees between the shadows related to the distance the earth moves?

9. From your measurements, how far did the earth travel in 15 minutes? If this is different from the calculated value, why do you think there is a discrepancy? (Hint: think about the shape of the earth)

10. Looking at your shadows, where do you think the sun rose this morning?

11. Where do you think it will set?

12. Use the compass to find east. Compare the shadows you drew to the compass heading. Does the sun appear to rise in the absolute east? How do you know?

Radioactive Decay: A Dating Game!

Objectives:
    Students will graph data.
    Students will interpret data from a graph.
    Students will operationally define half-life.
    Students will state the relationship between radioactivity and energy.

Materials:
    Box with lid, 100 black-eyed peas, pencil

Procedure:
1. Mark the box on the inside by putting an X on two of the opposite walls.
2. Place the beans in the box.
3. Secure the lid and shake the box for a specific time. (Use the same method and time for later "shakes".)
4. Open the box and remove all the peas whose "eyes" face the marked sides of the box. These represent atoms of the element that have decayed.
5. Record the number of remaining "atoms" in the box. Repeat steps 3, 4, and 5 until all the atoms are gone.
6. Graph the data below.

Results:
1. Determine the half-life of your element (the point at which only 50 atoms are in the box).

2. What is the half-life of your neighbor's element?

3. If the number of "decays" were in years (that is, one decay = 100 or 1000 years), what would be the half-life of your radioactive element?

4. What is the reason for the **instructions** in step 3?

5. What is the average half-life for the class?

6. What does "half-life" mean?

7. How is a half-life important to a geologist? How is it important to a medical doctor?

8. How is radioactive decay important to the energy industry?

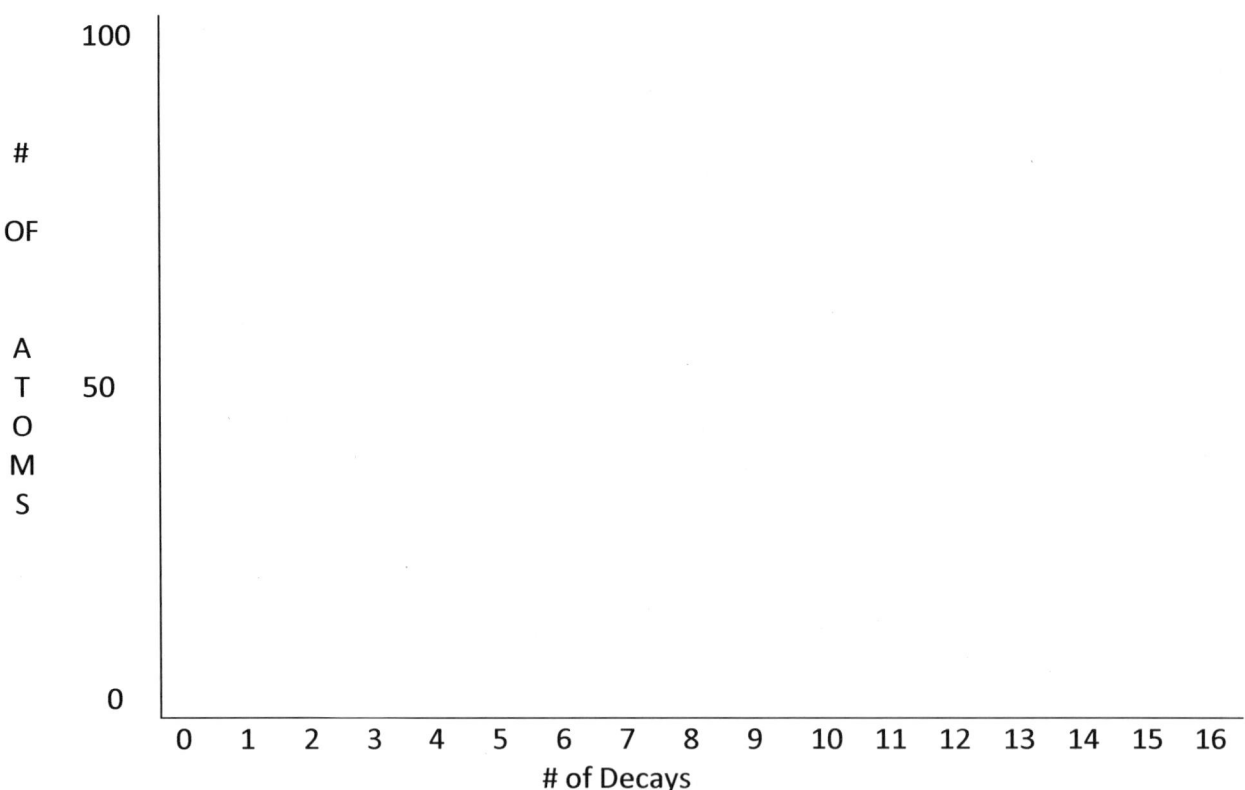

Simple Tests, Simple Answers

Objectives:
　Students will observe earth materials.
　Students will classify objects according to similarities and differences.

Materials:
　Samples with numbers, streak plates, knife, file, penny and fingernail, magnifying glass

Procedure and Results:
1. Test each unknown rock/mineral and complete the following chart.

Sample #	Hardness	Streak	Color	Luster	Texture

2. Which is the hardest? The softest?

3. Do any leave a streak? Which (if any)? What colors are the streaks?

4. What can you say about streak and color?

5. If you can, label the samples igneous, metamorphic and sedimentary, write the label I, M, or S beside the Sample #.

Using the "keys" on the following pages, identify the actual names of these samples.

Sample #	Name

6. Why is it important to know about rocks and minerals?

Metamorphic Rocks

Rock	Composition	Characteristics
Gneiss	Quartz, feldspar with mica, hornblende, or augite	Minerals arranged in coarse, irregular, parallel bands; resembles granite; formed from igneous rocks (granite, syenite)
Schist	Quartz with mica, hornblende or garnet	Minerals arranged in thin, regular, parallel bands; can be split; has wavy, uneven surface; formed from sedimentary (shales) or igneous rocks (feslites, basalts)
Slate	Clay minerals	Slaty cleavage; splits into smooth, flat surfaces; red or grey; formed from shale
Quartzite	Quartz sand	Very crystalline, particles hard, dense; formed from sandstone
Marble	Calcite (lime carbonate) or dolomite (magnesium carbonate)	Crystalline; granular; bubbles with cold HCl; white or grey streaked with colors due to impurities (iron, carbon). Dolomite marble only bubbles in heated HCl.
Anthracite coal	Plant remains (95% carbon)	Hard, shiny, shell-like fracture; doesn't crumble; from bituminous coal
Phyllite	Metamorphosed slate	Silky luster, due to fine grains of mica; silver grey

Igneous Rocks

If the rock is light colored, it is made mostly of light colored minerals (quartz and orthoclase feldspar). If it is dark colored, it is made mostly of hornblende, augite, olivine and plagioclase feldspar. The texture may be coarse-grained (individual minerals are visible), fine-grained (minerals are not visible), cellular (holey) or glassy.

Texture	Light Colored, Low Density	Dark Colored, High Density
Coarse-grained	Granite Syenite: no quartz, mostly orthoclase and one or more dark minerals	Gabbro: no quartz Diorite: some quartz; half light, half dark
Fine-grained	Felsite (Rhyolite and Trachyte): stony or flinty in appearance	Basalt: often has oval cavities filled with quartz, calcite, or olivine
Cellular	Pumice: frothy, spongy; silky appearance; porous, often floats on water	Scoria: resembles furnace cinders; similar to pumice, but darker and has larger holes
Glassy	Obsidian: shell-like fracture; usually black; sharp edges which are translucent	Basaltic glass or basaltic obsidian; thin edges are opaque

Sedimentary Rocks

Sediments	Hardening	Characteristics	Name
Plant remains	Compacted	Brown to black in color; soft; will burn	Coal
Shells, shell fragments, coral fragments, some limy mud	Calcite cement	Soft friable; crumbly. Soft, effervescent; may contain fossil shells	Chalk Limestone
Gravel	Calcite, silica, iron oxide, or clay	Rounded pebbles or boulders of rocks, held together by fine-grained materials.	Conglomerate
Sand	compaction and/or cement	Cemented sand grains; may be any color; with feldspar	Sandstone Arkose
Clay or Mud		Consolidated clay or mud with flat fracture; often has an earthy odor; any color	Shale
Limy Mud	Crystallized from solution	Will not scratch glass; not gritty; effervescent in weak acid; any color; may be fossilized	Limestone
Silica		Scratches glass; fine-grained; any color	Chert Opal Flint
Calcium sulfate plus water		Easily scratched with fingernail; usually white	Gypsum

Splish, Splash

Objectives:
Students will demonstrate the process of splash erosion.
Students will compare craters on earth to those formed on the moon.
Students will differentiate relative ages of craters.

Materials:
Flour, sand, small gravel (aquarium), dropper or pipette, water, small pie pans, newspaper and magnifying lens

Procedure:
1. Fill 1 pan with flour to a depth of 1cm. Make the surface smooth and place the pan in the center of the newspaper.
2. Using the dropper, let a large drop of water fall from 1 meter into the pan. Repeat this step several times, forming several craters. Record the results.
3. Dampen the flour and smooth the surface. Repeat the process and record the results.
4. Repeat these steps for the dry sand, wet sand, dry gravel and wet gravel. Record all results.

Results:

Material	Description of Craters
Dry Flour	
Wet Flour	
Dry Sand	
Wet Sand	
Dry Gravel	
Wet Gravel	

©2016 Near-Normal Design and Production Studio

Splish, Splash

1. Which material created the largest crater?

2. Which material would erode the quickest from rainwater? The slowest?

3. Compare the results of wet flour and wet sand with that of wet gravel. What were some of the differences?

4. When observed with the magnifying lens, how did the splash craters look?

5. Which of the materials you used is most like the surface of the moon? Why?

6. Which of these materials produce craters most like those on the moon? Why do you think this?

7. What would happen to moon craters if it rained?

8. How can you tell which craters were formed first?

©2016 Near-Normal Design and Production Studio

Sunrise, Sunset

Objectives:
  Students will estimate where the sun rises and sets.
  Students will measure the amount of rotation the earth makes in a given time period.
  Students will determine the relationship between the rotation of the earth and the rising/setting of the sun.

Materials:
  Large nail or tent stake, lined paper, sunny patch of ground, compass, marker

Procedure and Results:
1. Set the paper on the ground, using the compass to make the lines point due north. Poke the nail through the center of the paper. Make sure the shadow of the nail falls on the paper.
2. Use the marker to draw the shadow and label it "A".
3. Do not move the paper. Wait 15 minutes, repeat step 2, but label this shadow "B". How many lines on the paper are between shadows A and B?

4. Wait another 15 minutes, repeat step 2, labeling this shadow "C". How many lines are between A and C?

5. From your shadows, which way is the earth turning?

6. Which direction does the sun appear to rise?

7. Which direction does it appear to set?

8. If you left your paper out all day, what would the shadows look like?

©2016 Near-Normal Design and Production Studio

Sunrise, Sunset

9. Does it feel like the earth is moving? How do you know the earth is moving?

10. How is the rotation of the earth related to the rising and setting of the sun?

Swirls and Curls

Objectives:
   Students will differentiate clockwise and counterclockwise.
   Students will observe how clockwise and counterclockwise movement affects water.
   Students will infer from these observations how rotation of the earth affects weather and winds.

Materials:
   For each pair of students: 20cm diameter cardboard circle, pencil with a good eraser, thumbtack, clay, dropper or pipette, 25mls colored water, and 2 lab aprons

Procedure:
1. Push thumbtack through the exact center of the cardboard and into the pencil eraser.
2. Mount the pencil in the clay making sure that the pencil stands by itself in an upright position.
3. Make sure the cardboard spins freely.

Results:
1. The cardboard disk is the earth as you would see it from the North Pole. Spin the cardboard to the left (counterclockwise). What part of the earth did the cardboard represent when you turned it counterclockwise?

2. As the cardboard is spinning, quickly squirt a steady stream of water in a straight line from the center to the edge of the cardboard. What did the path of the water do?

3. If the wind starts from the North Pole, what happens to it as it goes toward the equator?

4. Turn the cardboard over and do the experiment again, but now pretend that you are at the South Pole. Spin the cardboard clockwise (to the right). Now what part of the earth does the cardboard represent?

©2016 Near-Normal Design and Production Studio

5. Observe the path of the water. What path did the water follow?

6. If the wind starts form the South Pole, what happens to it as it goes toward the equator?

7. Are different parts of the earth spinning in opposite directions? Why did you have to spin the cardboard in opposite directions?

8. If there were no other forces on that water in the sink, how would the Coriolis effect cause the water to behave?

9. How would the water in your sink behave if you lived in Australia and there were no other forces on that water?

Temperature: The Oceans' Pathfinder

Objectives:
   Students will manipulate laboratory equipment.
   Students will describe objects from their environment.
   Students will relate their observations to maps of the ocean currents.

Materials:
   Hot and cold water, beakers, food coloring, droppers, map of ocean currents

Procedure:
1. Pour about 200ml of hot water into one beaker and 200ml of cold water into another.
2. Place a single drop of food coloring in each of the beakers.

Results:
1. What happened to the food color in the cold water?

2. What happened to the food color in the warm water?

3. Were your observations different for the two beakers? Why?

4. Look at the map of world ocean currents. What can you say about the direction of flow of the Peru Current?

5. What about the Gulf Stream current's direction?

Temperature: The Oceans' Pathfinder

6. How do your observations of the cold water relate to your answer to question #4?

7. How do your observations of the warm water relate to your answer to question #5?

8. Why is it important for the currents to circulate around the earth?

Temperature: The Oceans' Pathfinder

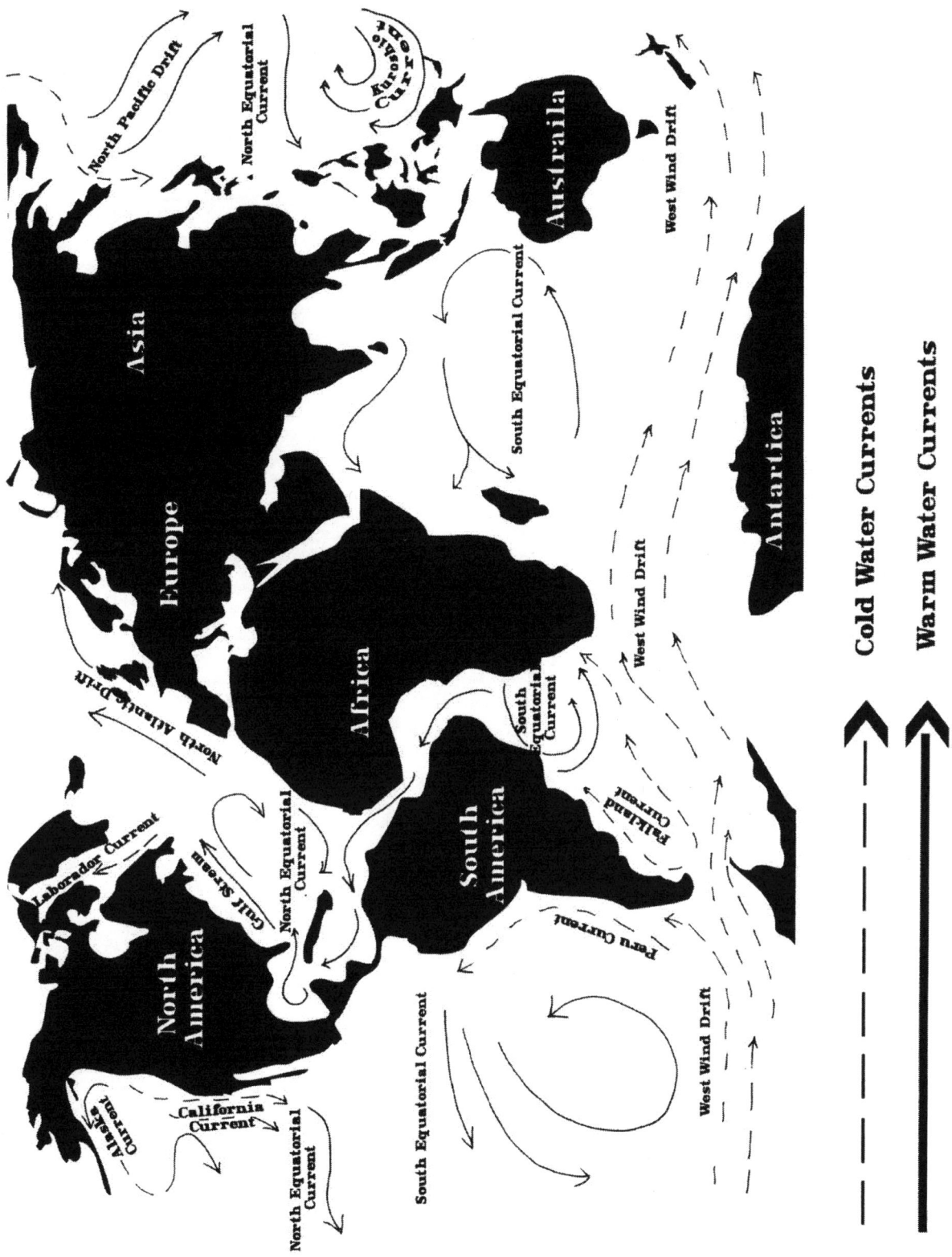

Them Bones, Them Bones!

Objectives:
Students will identify characteristics through inference.
Students will hypothesize the possible diet, environment, and mobility of the dinosaur.

Materials:
Scissors, skeleton master sheets, tape

Procedure:
1. "Chisel" (cut) the newly discovered bones in smaller sections for handling and labeling.
2. Roughly arrange them into the appropriate order.
3. When you have time you can chisel the bones out exactly and use the tape to construct the fossil.

Results:
1. Where do you think this dinosaur lived (land, water, air)? How do you know?

2. Was this dinosaur a plant-eater? How do you know?

3. Could this beast run very fast? How do you know?

4. Why did it have such a long tail?

5. What can you say about the environment in which this creature lived?

6. How do you know if all the bones are there?

7. How do paleontologists put together dinosaurs if they don't have all the bones?

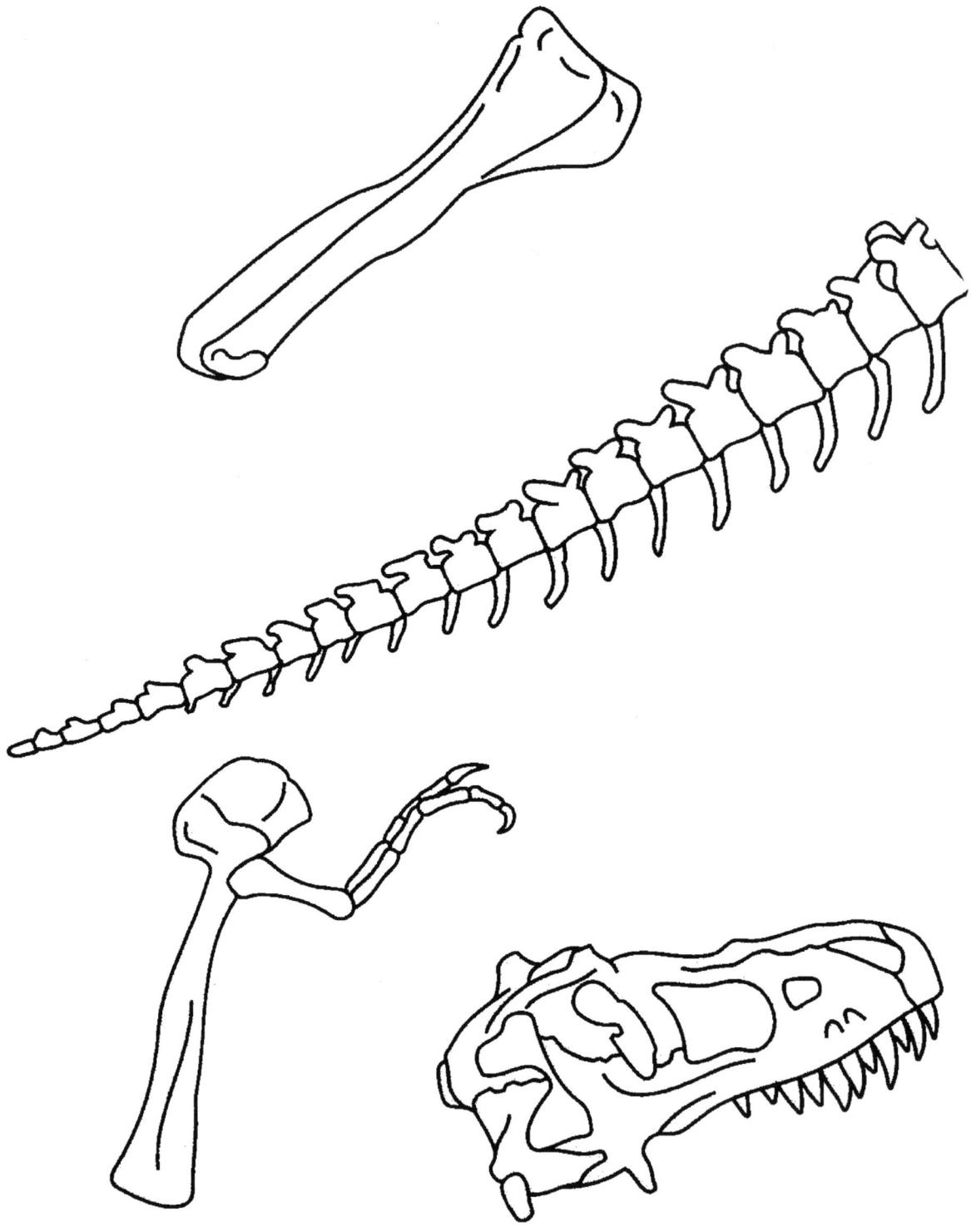

Them Bones, Them Bones!

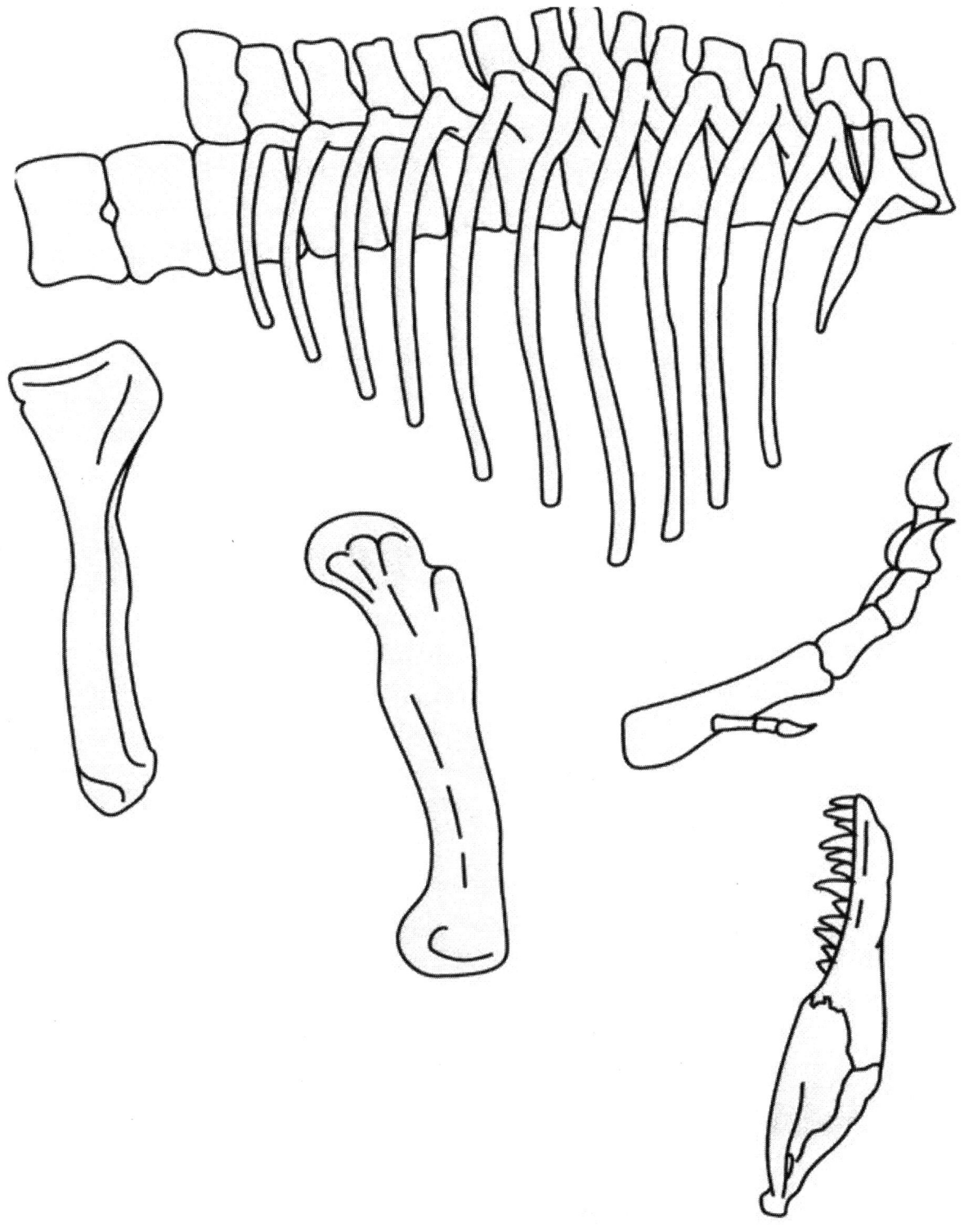

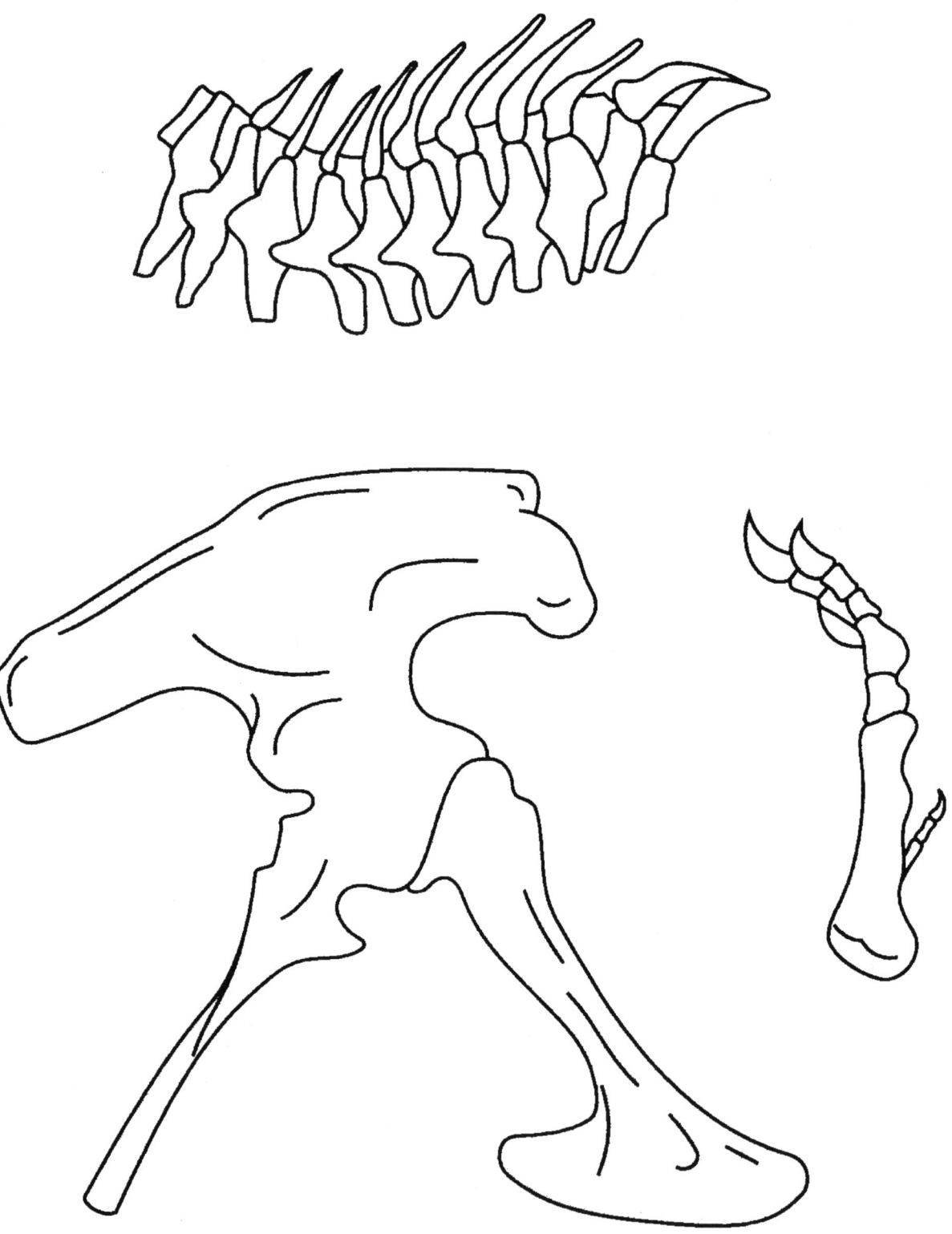

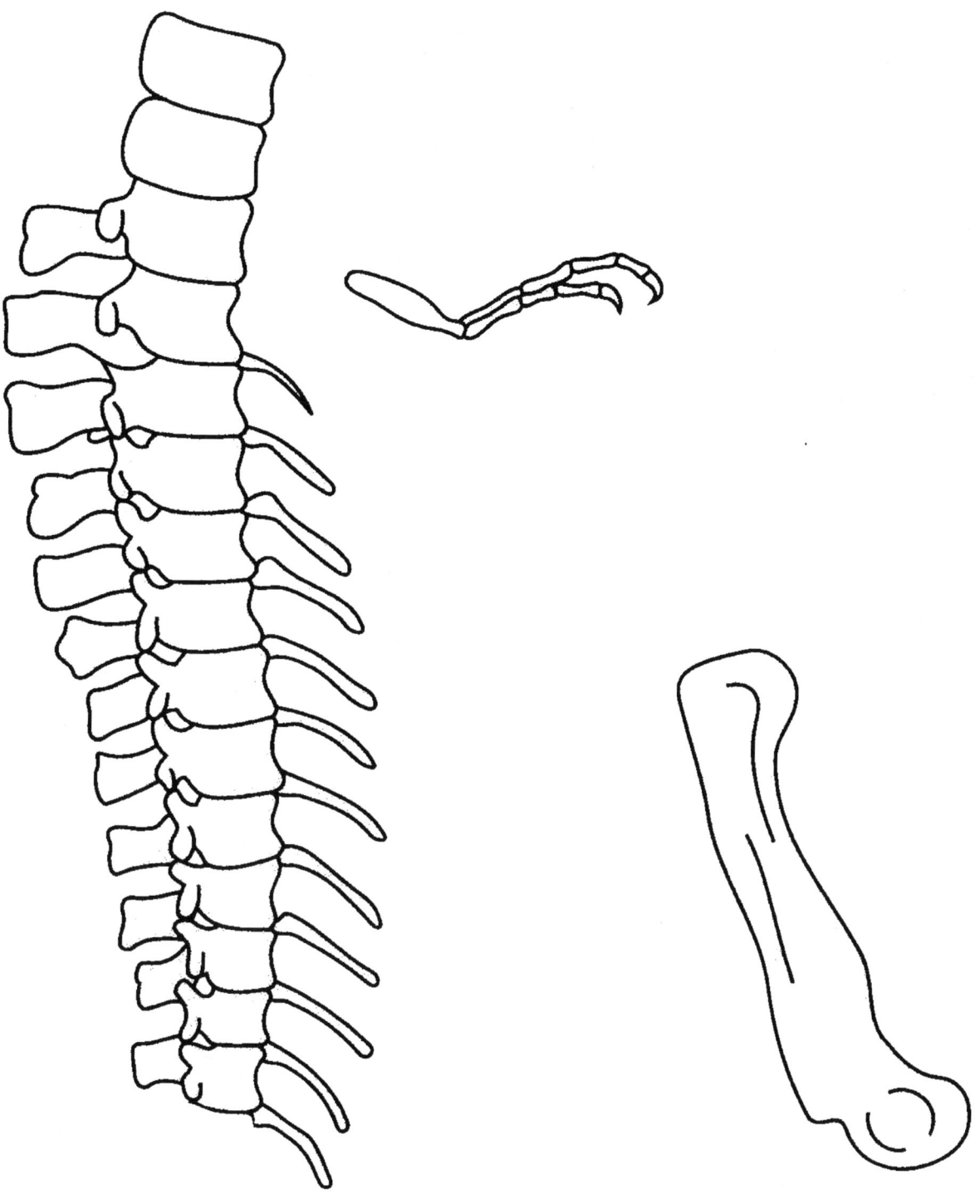

Trains, Race Cars, and Sirens

Objectives:
   Students will operationally define the Doppler Effect.
   Students will give examples of objects that produce the Doppler Effect.
   Students will state the relationship between speed and sound waves.

Materials:
   Large outside area, loud students

Procedure:
1. Go outside to an area approximately 25 meters square. This area should preferably be away from other classrooms.
2. One student stands in the center of this area.
3. A second student runs past the first, screaming.
4. Reverse roles and repeat the experiment.

Results:
1. What did you hear as the person ran toward you?

2. What happened to the sound as the person passed you?

3. What happened to the sound waves during this experiment?

4. Where have you heard this before (give 4 examples)?

5. What is happening to the sound waves as the person runs toward you? Away from you?

6. If sound waves were visible in this experiment, what would they look like as the person ran toward you? As the person ran away from you?

Water, Water, Everywhere ... But How Much?

Objectives:
Students will predict amounts of water from data supplied.
Students will relate available water to resource management.
Students will relate water usage to renewable resources.

Materials:
6 small containers, dropper and water supply

Procedure:
Large amounts of water are part of our everyday lives. Oceans, seas, lakes, rivers, ponds, and streams are familiar to everyone. Snow, rain, and hail are common to most people, and few can say that they have never been in a thunderstorm. But how is the total amount of water on the earth distributed?

1. You have six containers. One is labeled "Total Supply" and should contain 100 drops of water. Count that many drops into the container. Each drop is proportional to one percent of the earth's water. The other containers are labeled "Glaciers", "Atmosphere", "Ground Water", "Oceans" and "Rivers and Lakes".
2. Your task is to remove each drop from the total supply and place it in one of the containers, so that when you are finished you will have an accurate representation of the world distribution of water. There will be no drops left in the "Total Supply", all the water will be in the other containers.
3. When you have decided how much should go where, fill in the buckets below to illustrate your choices and give the percent you assigned to each category.

Results:

Glaciers	Rivers & Lakes	Air	Ground Water	Oceans
____ %	____ %	____ %	____ %	____ %

©2016 Near-Normal Design and Production Studio

1. Actual Percentages

   _____% _____% _____% _____% _____%

2. In which areas were your guesses closest?

3. Which of these locations are water we use for drinking?

4. Why are these the only sources of fresh water?

5. How do you think modern technology affects our water supply?

6. What steps must we take to protect these supplies?

7. What are some methods that could be used to make fresh water more accessible?

8. Although water is a renewable resource, why should care be given to avoid polluting any water source?

Teachers' Notes

Objectives:
   Students will predict amounts of water from data supplied.
   Students will relate available water to resource management.
   Students will relate water usage to renewable resources.

Materials:
   6 small containers, dropper and water supply

The following are the actual percentages of water in the listed locations. You may want the students to put their data on an overhead transparency or a chalk board prior to revealing the actual numbers. This allows students to confront their misconceptions.

Students may also perform this activity using 1000 drops in the "Total Supply" rather than 100 drops. This changes the math slightly, but allows more accurate allocation of resources.

Glaciers - 2.0 %

Rivers and Lakes - 0.01 %

Air - 0.001 %

Ground water - 0.5 %

Oceans - 97.5 %

Where Do I Get My Energy?
Teacher's Instructions

Objectives:
    Students will match the energy type to the pictures.
    Students will hypothesize which of the objects use renewable energy and which use non-renewable energy.
    Students will determine what practices would save energy.

Materials:
    Handout, pencil, kite, light bulb, snail (or other animal), battery powered toy car, television, plant, toy sail boat in pan of water, matches (for teacher demonstration only), plant (to represent tree)

Procedure and Results to Discuss:

1. Show the students the items shown on their handout. Ask them what type of energy each uses.

2. Allow students to touch all the items except the matches.

3. Once students have made their hypotheses, ask them to take turns making the object use energy (except the matches, the animal and the plant).

4. Ask students if a real car uses the same kind of energy as a toy car.

5. Discuss the types of energy animals and plants use and where the energy is produced.

6. Strike a match and discuss the burning of wood.

7. Once you are satisfied that the students understand what type of energy each object uses, ask them to draw a line connecting the energy words to the pictures.

8. Ask them which forms of energy are renewable and which are not.

9. Ask students what they would do to conserve nonrenewable energy.

©2016 Near-Normal Design and Production Studio

Where Do I Get My Energy?

Draw a line from the energy word to the picture of the object that uses that kind of energy.

Gasoline	Sun	Wind	Wood
Electricity	Wind	Electricity	Food

Life Sciences

Adaptation and Survival

Objectives:
    Students will identify some of the factors leading to adaptations.
    Students will describe the changes in a population.
    Students will research information regarding the Pepper Moth.
    Students will draw conclusions from data.

Materials for each pair of students:
    2 sheets of newspaper, scissors, 100 moths, clock with a second hand or stop watch, library access or Internet access

Procedure:
1. Using one of the sheets of newspaper, cut out 100 moths (see pattern).
2. Lay the other sheet of newspaper on the floor.
3. While your partner is not looking, sprinkle 50 moths on the newspaper. This is the first generation of moths.
4. Your partner is now a bird. From a standing position, he or she has 10 seconds to pick up as many moths as possible by repeatedly swooping down, and standing up. **Moths must be caught one at a time.**
5. At the end of 10 seconds, count the number of moths collected and record this number on the chart below. Subtract this number from the beginning number. This is the number of survivors.
6. Add as many new moths to the newspaper as there are survivors. (The survivors have reproduced forming the second generation.)
7. Repeat steps 4 through 6 three more times for each partner for a total of four generations each.

Results:

Generation	# Eaten	# of Survivors	# of Offspring
1			
2			
3			
4			
5			
6			
7			
8			

Adaptation and Survival

1. Look at the number of moths in each generation. What can you say about the trend in this data?

2. What is the difference in coloration between the moths eaten and those that survived?

3. What would have happened to the moths if they were not adapted to their environment?

4. What would happen to the birds that ate the moths if the moths became extinct?

5. What would happen in Generation 8 if you used the Sunday comics as background for Generation 7?

6. How have humans changed physically in the last 500 years?

7. Will adaptation of the human race be important in the next 200 years? How and why?

8. What does this activity have to do with the Pepper Moths in England?

©2016 Near-Normal Design and Production Studio

Air Pollution

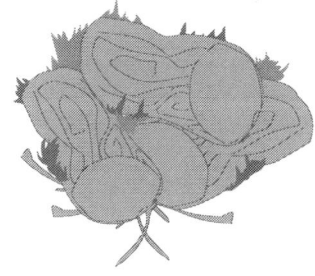

Objectives:
Students will determine whether the air contains large amounts of dust or other particles.
Students will differentiate between pollution from human activity (soot, building materials, etc.) and naturally occurring pollution (pollen, fungus, etc.).

Materials:
Double sided tape, 4 microscope slides, hand lenses

Procedures:
1. Put the tape on one side of the slide. Cover about three-fourths of the slide.
2. Place the slides in different areas of your school. Leave them there for a week.
3. Collect the slides and label them with their location. Look at the slides with a hand lens. On the data table, write whether the particles you see are dust, plant material, fibers, or other types of materials. Draw and describe the "other types of materials" you see.

Results:

Location	Type of Material(s)	Description/Drawing

1. Which slide had the most particles? Which had the least?

2. Which of the materials might have come from human activity? How can you tell?

3. What can you say about the pollution on the slide with the most particles?

4. Do you think any of these particles might be harmful to humans or plants? Which ones?

5. Predict how your life would change if there were twice as many particles on the slide.

6. How could air pollution in your school be decreased?

Alive Or ???
Teacher's Instructions

Objectives:
    Students will classify objects from the environment as living, once living, or non-living.
    Students will give reasons for their choice of classification schemes.
    Students will provide examples of items that fit into the categories.

Materials:
    Microscope, hand lens, pipette or dropper, culture dish, microscope slide, plant, small animal, pond water with active protists, rocks, center area, tags labeled "living", non-living", "once living"

Procedure and Results to Discuss:
1. Prepare the microscope by putting a drop of pond water on the slide, putting the slide on the microscope stage, and adjusting the focus. If you have a projecting microscope available, you may want to use it for this portion of the activity.

2. Allow the students to examine each of the other materials listed above. Remind them of proper handling of the plant and the animal.

3. Ask students which of the materials are living or non-living. Ask them to explain their answers. Suggest that some items were once living, but now are not. Ask them if any of the objects they see fit into this category.

4. Place labels "Living", "Non-Living", and "Once Living" on the Center table. Ask students come to the Center to place these labels on the objects they feel fit into these categories.

5. Ask students to give examples of other things that fit into these categories.

©2016 Near-Normal Design and Production Studio

The Beat Goes On: Part 1

Objectives:
    Students will observe natural phenomena.
    Students will identify parts of the circulatory system.
    Students will measure pulse rates.

Materials:
    Long cocktail toothpick or half a straw, dime, clay

Procedure:
1. Mash a small amount of clay onto the dime, and position the toothpick in it so that it stands upright.
2. Place your forearm on a desk with your palm up.
3. Place the dime over the blood vessels in the wrist. Observe the toothpick. If nothing happens, move the dime around until the toothpick starts to move.

Results:
1. What causes the toothpick to move?

2. How many times does the toothpick moves in 6 seconds?

3. Multiple the number you counted by 10.

4. What is this number called?

5. Why would it be important to be able to monitor the movement of the toothpick?

6. What system of the body are you observing? How do you know?

©2016 Near-Normal Design and Production Studio

The Beat Goes On: Part 2

Objectives:
    Students will observe natural phenomena.
    Students will identify parts of the circulatory system.
    Students will measure pulse rates in given situations.
    Students will predict heart rates based on health and fitness.

Materials:
    Long cocktail toothpick or half a straw, dime, clay

Procedure:
1. Mash a small amount of clay onto the dime, and position the toothpick in it so that it stands upright.
2. Place the dime over the blood vessels in the wrist. Observe the toothpick. If nothing happens, move the dime around until the toothpick starts to move.
3. Count the number of times the toothpick moves in 6 seconds. Record this number in the Results section.
4. Jog in place or do some other form of exercise for about 2 minutes.
5. Place the dime over the blood vessels in the wrist. Observe the toothpick. If nothing happens, move the dime around until the toothpick starts to move.
6. Count the number of times the toothpick moves in 6 seconds. Record this number in the Results section.

Results:
1. How many times did the toothpick move in 6 seconds before jogging? Multiple this number by 10.

2. What happened to the movement of the toothpick after jogging? Record the rate. Multiple this number by 10.

3. Why has this happened?

4. What would happen if you exercised for a longer time?

©2016 Near-Normal Design and Production Studio

The Beat Goes On:
Part 2

5. Check your pulse rate again. What is it now (remember to monitor for 6 seconds then multiply by 10)?

6. Why is it important for athletes to monitor their pulse rates?

7. Would you expect an athlete's pulse rate to be faster or slower than a "couch potato's" after jogging? Why?

8. How is health related to the circulatory system?

9. What parts of the circulatory system have you been able to indirectly observe in this activity?

Blink or Wink

Objectives:
   Students will distinguish between voluntary muscles and involuntary muscles.
   Students will give examples of both involuntary and voluntary muscles.
   Students will analyze data.

Materials:
   Watch or clock with second hand

Procedure:
   Muscles may be classified into two groups: voluntary and involuntary. Voluntary muscles are those that you can control. Involuntary muscles are those you cannot control. Some involuntary muscles may be controlled for a short period of time.

1. Blink your eyelids three times. Now try not to blink. Have your partner time you and record the data.
2. Repeat two more times and record the times in the chart below.
3. Now have your partner repeat steps 1 and 2.
4. Average your times (total/3). Record these numbers in the data table.

Results:

Trial	Your Time in Seconds	Partner's Time in Seconds
1		
2		
3		
Total		
Average		

1. Compare your data with the rest of the class. What was the longest average?

2. What was the shortest average?

3. What was the most common average?

4. From your data, how long would you expect an average person to be able to control his/her eyelids?

5. Why are your eyelids not completely controllable (voluntary)?

6. What other involuntary muscles can you control for a short time?

7. List examples of 5 voluntary muscles.

8. There are involuntary muscles you cannot control even for a short period. List 5 of them.

9. What happens to voluntary muscles when you sleep? How do you know?

10. What happens to involuntary muscles when you sleep? How do you know?

Designer Genes!

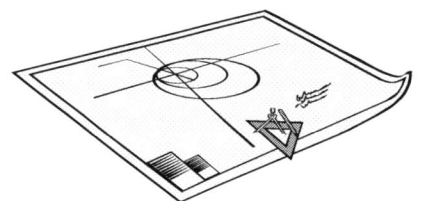

Objectives:
　Students will classify objects or events according to similarities and differences.
　Students will operationally define dominant and recessive.

Materials:
　20 red and 20 white beads, two wide-mouthed plastic jars, beakers or paper lunch bags

Procedures:
　Gregor Mendel observed inherited traits or characteristics that living things share. He wondered why certain traits found in parents show up in their offspring while others did not. In his research he found these traits are passed on but some are recessive and others dominant in offspring.

1. Place 10 red beads and 10 white beads into each of two jars (20 beads per jar).
2. Shake each jar. Close your eyes and pick one bead from each jar. Repeat until all beads have been selected.

　Red = Dominant (R)　White = Recessive (r)

　R + R = Red offspring (pure)
　R + r = Red offspring (hybrid)
　r + R = Red offspring (hybrid)
　r + r = White offspring (pure)

Results:

Trial #	Jar 1	Jar 2	Offspring Trait	Trail #	Jar 1	Jar 2	Offspring Trait
1				11			
2				12			
3				13			
4				14			
5				15			
6				16			
7				17			
8				18			
9				19			
10				20			

©2016 Near-Normal Design and Production Studio

1. Why did you use two jars for this lab?

2. How many red offspring did you get?

3. How many white offspring did you get?

4. What was the ratio of red to white? What was the ratio of red (RR) to red (Rr) to white (rr)? What do these ratios mean?

5. What are you chances of getting a white offspring?

6. Why would you want to know the likelihood of getting certain traits in the offspring before you crossbred wheat plants?

Feast or Famine?

Objectives:
    Students will observe and measure plant growth.
    Students will evaluate how variables such as sunlight and water affect plant growth.
    Students will create an experiment that tests a hypothesis about plant growth.

Materials:
    2 foam cups, 4 bean or corn seeds

Procedure and Results:

1. You are to make up an experiment that will test a hypothesis about growing plants. You will need to keep a daily log of your observations and manipulations of the plants. One cup is to be your control. The other cup is to be your experiment. Remember, the dependent variable is how much your plants grew. The manipulated variable can be of your choosing (amount of light, water, place in the house or outside, etc.). The control and the experimental plants should be treated the same, except for **one** variable.

2. Write your hypothesis. (If I do _____ to the plant then _____ will happen.)

3. Describe your experiment.

4. What is your manipulated variable?

5. Find some dirt near your school or home. Plant the seeds. On another sheet of paper (or in a notebook) keep a record of your observations, what you did, and how the plants reacted.

©2016 Near-Normal Design and Production Studio

In two weeks answer the following questions from the data you collected:

6. What were the results of your experiment?

7. Do your results support your hypothesis? Why or why not?

8. If you could do your experiment again, is there anything you would change? Explain.

9. What did you learn from your experiment that you can use in other experiments?

Fuzzy or Furry

Objectives:
   Students will make observations and inferences regarding a habitat.
   Students will state relationships between biotic and abiotic portions of the environment.

Materials:
   8 large dark fuzzies, 16 small dark fuzzies, 8 large light fuzzies, 16 small light fuzzies, scissors, area with shrubs and grass

Procedures:
1. Divide into twos or threes. Go outside to the habitat area with your teacher.
2. Make a list of observations about the habitat and its inhabitants.
3. Also list inferences you make from your observations.

Results:
1. Define observation.

2. Define inference.

3. Make a list of 10 observations.

4. Make a list of 10 inferences.

5. How must animals interact with their environment to be able to survive?

6. How is this model of a habitat different from an actual habitat? How is it like a real habitat?

7. Describe your habitat.

8. How is your habitat like this model habitat? How is it different?

Fuzzy or Furry
Teacher's Instructions

Objectives:
    Students will make observations and inferences regarding a habitat.
    Students will state relationships between biotic and abiotic portions of the environment.

Materials:
    8 large dark fuzzies, 16 small dark fuzzies, 8 large light fuzzies, 16 small light fuzzies, scissors, area with shrubs and grass, glue, eyes (optional), scissors, craft fur or small and large pom-poms

1. Cut fuzzies from craft fur. You may want to add glue-on eyes.
2. The larger ones represent the adults, the smaller are the offspring.
3. The following is a suggested scenario; be creative. Females are never alone with offspring. Males die unless they are with males or offspring. Offspring are always found in even numbers. Offspring can be with both parents, but must be with a male. Females may be alone or in female groups. Rarely are any animals seen in open areas; they congregate near trees and shrubs. Their diet may include acorns, white rocks, etc. and is found near the congregation. Males can be dark and females light, or vice-versa. Try to stay away from stereotypes.

Hot or Cold?

Objectives:
 Students will measure the effects of hot/cold (inside/outside) temperatures on microbes.
 Students will form hypotheses.
 Students will evaluate data and draw conclusions.
 Students will observe the variety of microbes present in the air.

Materials:
 Petri dishes with culture medium, thermometers and hand lenses

Procedure:
1. Expose the culture mediums to the air for 10 minutes, re-cover and place one dish outside (or in a warm area of the room) with a thermometer and the other dish inside (or in a cool area of the room) with a thermometer.
2. Write hypotheses concerning what you think will grow (if anything) on the culture, and one about the effect of temperature on the growth.

Results:
1. Hypothesis 1 (will anything grow):

2. Hypothesis 2 (if something grows, will it do better in the cooler environment or warmer environment):

3. Day 1: Examine the petri dishes with both the unaided eye and with the hand lens. Draw what you see in the circle below. Why do you need to make this drawing?

4. Examine the culture dishes each day for the next three days and draw what you see.

	Day 1	Day 2	Day 3	Day 4
Inside				
Outside				

5. What happened to the culture growths?

6. Which grew the most? Inside? Outside?

7. What can you say about the effect of temperature on culture growth?

Hot or Cold?
Teacher's Instructions

Objectives:
   Students will measure the effect of hot/cold (inside/outside) temperatures on microbes.
   Students will form hypotheses.
   Students will evaluate data and draw conclusions.
   Students will observe the variety of microbes present in the air.

Materials:
   Petri dishes with culture medium, thermometers and hand lenses, unflavored gelatin, chicken bullion

Procedure:
1. Follow the label directions for preparing the gelatin, but dissolve one chicken bouillon cube in slightly less hot water than directed.
2. Open the first petri dish and fill it about ¼ full of the hot gelatin. Immediately close the petri dish. Your goal is to insure that the culture medium is as sterile as possible.
3. Continue filling each dish. The gelatin should "set-up" without having to be refrigerated.
4. Encourage students to select warm places out of direct sunlight.
5. Most likely, each dish will have a nice growth of fungus. The fungi may cause the gelatin to melt. Although the lids may be removed, **caution students** to avoid touching the fungi in their dishes.

©2016 Near-Normal Design and Production Studio

Keep in Touch

Objectives:
    Students will differentiate between reflex and reaction times.
    Students will indirectly observe the nervous system.

Materials: Pencil, paper, stopwatches

Procedure and Results:
1. Divide students into two groups. Two students will be the timers/recorders.
2. Students in each group will line up one behind the other with their right hands slightly above the person's shoulder in front of them. The first person in line will hold his or her hand just above shoulder level.
3. When the teacher says, "Go!" the last student in line will tap the shoulder of the next person in line. This procedure continues until the person at the front of the line drops his or her hand.
4. The timers/reporters stop the watches and record the times.
5. Repeat this five times and record the times.

Trials	1	2	3	4	5
Times Group 1					
Times Group 2					

6. Follow the same procedure again except students hold both hands above shoulders of the person next in line. The persons at the head of the lines hold out both hands.
7. When the teacher says, "Go" the first person taps the right shoulder of the next person, who taps the left shoulder of the next, and so forth. Alternate taps throughout the line. The persons in front must drop the alternate hand from the shoulder tapped.
8. Repeat this five times and record the times.

Trials	1	2	3	4	5
Times Group 1					
Times Group 2					

9. Average of the first five trials.

10. Average of the second five trials.

©2016 Near-Normal Design and Production Studio

11. Why is the first average smaller than the second?

12. What is the difference between a reflex and a reaction? How do you know?

13. What parts of the nervous system are you indirectly observing?

14. Who should have a faster reaction time, a very tall athlete or a short athlete? Why?

Make a Place

Objectives:
    Students will construct a habitat.
    Students will predict the needs of a hypothetical animal.
    Students will state the relationships between the biotic and abiotic factors in an environment.

Materials:
    Pencil, paper, and map colors

Procedure:
    Scientists have discovered a new animal. Its picture is below. Your job is to determine what kind of zoo habitat would be best for Flybelius humpis.

What kind of food do you think it eats? How do you know?

In what kind of climate do you feel it will be most comfortable? Why?

What colors might an animal like this use for camouflage?

What must the environment provide for the Flybelius humpis to be healthy?

What was the natural habitat of Flybelius humpis like? How do you know?

Color your Flybelius humpis and write a paragraph about its zoo habitat.

More Designer Genes

Objectives:
   Students will classify objects according to similarities and differences.
   Students will make a model of a set of chromosomes.
   Students will observe similarities and differences.

Materials:
   Paper, scissors, pencil and tape

Procedure:
   A human body cell has 23 pairs of chromosomes in its nucleus. The two members of each pair of chromosomes have a matching size and shape. One member of each pair is from the person's father, and the other from his/her mother. A human male has an unmatched pair made up of an "X" and a "Y" chromosome.

1. On the following page, you will find a set of human chromosomes numbered 1 to 23 and lettered A to W.
2. Carefully cut out the chromosomes with the identifying number or letter.
3. Use the table for matching your chromosomes.
4. Arrange them from largest to smallest.

Results:
1. What factors did you use to match the chromosomes?

2. Is this is a male or female? How can you tell?

3. Which of the genes can you tell for certain came from the father? How do you know?

4. Why is it important to study genetics?

©2016 Near-Normal Design and Production Studio

More Designer Genes

10	H	16	R	I	12	W	1
Q	22	8	B	K	23	3	V
21	O	M	4	S	15	D	A
19	P	5	17	18	20	6	
T	14	F	C	J	L	G	
11	7	9	N	U	E	13	

More Designer Genes

Chromosome Pairs

Numbered	Lettered	Numbered	Lettered

Name Game

Objectives:
  Students will construct a name based on given data.
  Students will construct a 2-dimensional model based on given data.
  Students will interpret data.
  Students will analyze environmental interactions when constructing their model.

Materials:
  Paper, pencil and map colors

Procedure:
  You are a naturalist working for a zoo. Some explorers have captured a previously unknown herbivore. They sent you a description of its habitat but not of the animal itself. Your job is to name it using only the information you have about its habitat. The following data are available about its habitat:
- It rains two to three inches per day.
- There are very tall trees with the animals' favorite food located in the tops of the trees.
- The food is a nut with a very hard shell.
- The ground under the trees is very swampy.
- It is cool, about 65° F.
- Carnivores roam the forest floor.

Results:
1. How big is the animal?

2. How does it move around its habitat?

3. What color is it?

4. How does it eat its favorite food?

Name Game

5. How does it avoid its predators?

6. How is it adapted to deal with the wet climate?

7. What is a good scientific name for this creature?

8. Draw a picture of the animal and color it.

Presto changO!

Objectives:
Students will explain crossbreeding using an orange, a tangerine and a tangelo.
Students will relate genotype to phenotype.

Materials:
Orange, tangerine, tangelo, dull knife, pencil, paper, beaker, graduated cylinder and metric ruler

Procedure and Results:
1. Look at the specimens closely. List some of the characteristics of each.

	Orange	Tangerine	Tangelo
Color			
Size			
Shape			
Smell			

2. Peel each specimen and examine the sections and peeling. You may want to use the ruler for more precise measurements.

	Orange	Tangerine	Tangelo
Thickness of Peel			
Section Size			
Section Shape			

©2016 Near-Normal Design and Production Studio

3. Examine one section of each fruit. Be sure to make accurate measurements.

	Orange	Tangerine	Tangelo
Amount of Juice			
Number of Seeds			
Seed Size			
Taste			

4. From the data you have collected, describe the specimens.

5. Which of the specimens seems to be a mixture of the other two? Explain.

6. Are these specimens genetically different? How do you know?

7. Why do you think we crossbreed plants?

8. Are there other organisms that are crossbred? Give some examples.

The Rag Man

Objectives:
　　Students will separate, measure, and estimate the amount of solid waste produced by one class.
　　Students will relate energy to recycling.
　　Students will research the effects of solid waste on the environment.

Materials:
　　Bathroom scale, trash bags, solid wastes, plastic gloves, graph paper, access to the Internet or the library

Procedure:
1. Ten students will collect the solid waste from assigned classes as instructed by the teacher (near the beginning of the period).
2. In teams of 2 or 3, separate the trash from each class as thoroughly as possible.
3. Weigh each category of waste (paper, metal, and glass) and record the weights.
4. Calculate how much trash one class would generate in 180 days.
5. Calculate the amount of trash generated by the entire school in one day.
6. Calculate the amount of trash generated by the entire school in 180 days.
7. Compare your calculations with your classmates.
8. Find out how much energy it takes to recycle paper, metal and glass by doing research in the library or on the Internet.
9. Find out where solid waste that is not recycled is stored by doing research in the library or on the Internet.

Results:

	Paper	Metal	Glass	Other
Weights				
1 Class/day				
1 Class/180 days				
School/day				
School/180 days				

1. Which category was the largest? Which was the smallest?

2. How did your data compare with that of your classmates?

3. Why are your calculations different?

4. What can you conclude from this experiment?

5. What additional data could you collect?

6. How much energy does it take to recycle paper? How much for metal? How much for glass?

7. Where does the trash that your school does not recycle go?

8. How does recycling affect the environment? Think about both the positive and negative aspects.

9. How does dumping non-recycled materials affect the environment? What are the good and bad points?

10. What can your school do to reduce the amount of waste?

Right Eye, Left Eye, Which Eye?

Objectives:
 Students will compare the differing results.
 Students will determine which eye is dominant.
 Students will indirectly observe the nervous system.

Materials:
 Paper towel tube

Procedure and Results:
1. Hold the paper towel tube with both hands extended in front of you.
2. With both eyes open, sight through the tube toward some object across the room (such as a wall clock).
3. Now, while focused on the object across the room, close the left eye. What happened?

4. Repeat step #3, but this time close the right eye. What happened?

5. How do you explain the results from steps #3 and #4?

6. In your classroom which eye was dominant (the one focused on the object) most often?

Males		Females	
Left Eye	Right Eye	Left Eye	Right Eye

7. How did the dominant eyes compare with your classmates who are right-handed/left-handed?

Males		Females	
Left Hand	Right Hand	Left Hand	Right Hand

©2016 Near-Normal Design and Production Studio

8. Are there any individuals who are left-handed, but have a right dominant eye? What about the reverse?

Males		Females	
Left Hand/Right Eye	Right Hand/Left Eye	Left Hand/Right Eye	Right Hand/Left Eye

9. What is happening to the nerve impulses in the brain if a person is left-handed and right eye dominant?

10. What is the benefit of having one eye or hand dominant?

11. What does the nervous system have to do with this activity?

Sherlock Who?

Objectives:
    Students will identify similarities and differences of living things.
    Students will analyze data.
    Students will draw conclusions from data.

Material:
    Soft-lead pencil, notebook paper, transparent tape (¾")

Procedure:
1. Make a smudged area on the notebook paper with your pencil. (This will be your "ink-pad".)
2. Rub each finger in the smudged area until the fingertip is covered with the pencil lead.
3. With your free hand (or your partner's help) place a piece of the tape over the fingertip. This will "lift" the print off the finger.
4. Place the taped fingerprint in the blocks below.

Results:

1. Compare your fingerprints to those below. Which type of fingerprints do you have?

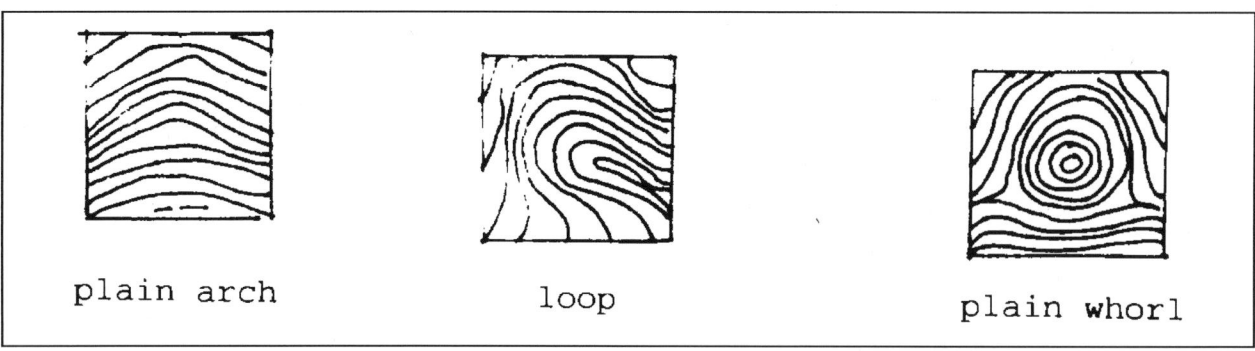

2. Are all your fingerprints just alike? Explain.

Sherlock Who?

3. If you found fingerprints on a stolen piece of candy or toy would you have enough evidence to know who the thief was? Why?

4. What additional information would be helpful?

5. How could fingerprints be more distinctive?

6. Which type of **index** fingerprints do your classmates have? (Use tic marks in the table.)

Plain Arch	Loop	Plain Whorl

7. From your data, what is the most common type of fingerprint?

8. How are scientists like detectives?

Smoke Signals

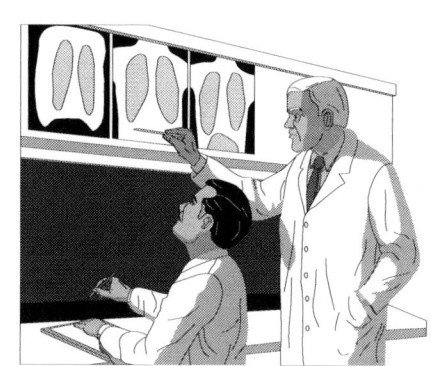

Objectives:
    Students will distinguish between responsible and irresponsible choices that affect personal health.
    Students will observe the materials produced by the burning of tobacco.

Materials:
    Test tube, test tube holder, burner, cotton-ball, water, ¼ cigarette for each lab group

Procedure:
    Chemicals, called carcinogens, are the cause of cancer. Tobacco is a carcinogen. It enters the body through smoking cigarettes, cigars, pipes, and also through breathing other peoples' smoke.

1. Place a few grams of tobacco into the test tube.
2. Loosely plug the top of the tube with the cotton wad.
3. Hold the tube over the burner as shown in the example below for about three minutes.
4. Cool the tube and remove the cotton ball.

Results:
1. Describe the appearance of the cotton wad after the tobacco has heated.

2. What does it feel like?

3. What does it smell like?

4. If this heated tobacco were breathed, where would this tar collect?

5. What happens if this tar gets into the body?

6. How does second-hand smoke affect children?

7. Why do some people smoke, even though they know it causes cancer?

Survival of the Fittest!

Objectives:
   Students will demonstrate how organisms adapt.
   Students will determine the effect of the environment on organisms.
   Students will predict changes that might occur in an environment.

Materials:
   Masking tape, peanuts with shells, plastic spoons, paper plates, paper, pencil, buttons on garments, shoes with laces

Procedure:
1. With a partner, tape your thumbs to the palms so that the thumbs cannot move and do not show above the tape.
2. Each person will perform the following tasks:
   - Open peanuts without using your thumbs. Eat several.
   - Use the plastic spoon and pick up 5 shelled peanuts to eat.
   - Write your name on a piece of paper.
   - Make a paper airplane.
   - Unbutton and re-button a garment.
   - Untie and re-tie a shoe.

Results:
1. What are you demonstrating in this exercise?

2. Give examples of how thumbs aid in food-getting.

3. How do thumbs assist an animal in protection in its environment?

4. How did you adapt, that is, obtain greater speed or skill in this activity?

5. What conclusion could you draw relative to an animal's ability to adapt?

6. Are humans still adapting or are we adapting our environment, or both? Explain.

7. If humans had no thumbs, how would our environment be different? What about our technology?

8. As people age, they may lose dexterity and strength in their hands. Describe some product changes that could help people with arthritis or loss of use of their hands.

Water Pollution

s:
nts will manipulate variables to determine the
of water from several sources.
Students will suggest reasons for differences in water samples.
Students will hypothesize on the levels of pollution found in drinking water.
Students will state methods for protecting water from pollution.

Materials:
Small bottle of liquid dishwasher soap, 4 test tubes, 20ml distilled water, 20ml water from drinking fountain, 20ml water from home, 20ml pond water, toothpicks, graduated cylinder, chalk

Procedure:
1. Label test tubes and fill them with samples of water from each source.
2. Add enough ground up chalk to the distilled water to turn the water milky. This is your "hard water" sample. Chalk is mostly calcium, one of the minerals that makes water "hard".
3. Using the toothpick, add a small amount of soap to each sample.
4. Cover securely and shake for 30 seconds.
5. Add more soap, drop at a time, until the bubbles last for 10 seconds after shaking.
6. Record the number of drops in the data table.

Results:

Locations	Test Tubes	# of Drops
Distilled	A	
Drinking Fountain	B	
Home	C	
Pond	D	
"Hard Water"	E	

1. Which is the manipulated (independent) variable in this experiment?

2. Which is the responding (dependent) variable?

Water Pollution

3. Compare your "hard water" to the other samples. How are they alike, and how are they different?

4. Which sample needed the most drops of soap?

5. Which sample (other than E) has the most minerals dissolved in it? How can you tell?

6. Which sample would be better for washing dishes? Why?

7. Should you use pond water in a steam iron? Why?

8. What is a possible result of adding waste water directly into the ground water?

What's in It?

Objectives:
    Students will determine which foods are appropriate for restrictive diets.
    Students will interpret nutritional value from labels.
    Students will compare "all-natural" products with common products.
    Students will use nutritional information to plan a meal.

Materials:
    Food labels, paper and pencil

Procedure:
1. Examine at least five (5) labels from food containers and list unfamiliar terms.

2. Complete the following chart of your products.

Name	Net Weight	Sugar	Salt	Calories	Protein	Vitamins

3. Which of these products would be "off-limits" if you were diabetic? Allergic to soybean? Needed a salt-free diet? Mark the "off-limits" products with an X.

Name	Sugar	Salt	Soybean

4. What is the difference between eating "all-natural" foods and those with additives? How do you know?

5. Why is it important for food labels to list all the ingredients?

6. Plan a meal for someone allergic to spinach and to onions. Make sure you include meats, fruits, and vegetables.

Where Did You Get It?

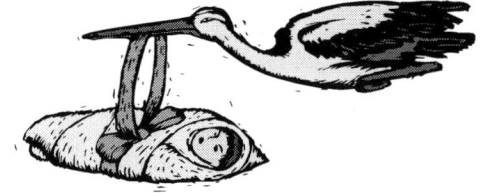

Objectives:
Students will determine the ratio of dominant genes to recessive genes.
Students will show how gene pairing occurs by chance.
Students will predict the possible phenotypes of an organism from parental genotypes.

Materials:
For each pair of students: 2 large containers (one labeled male, the other labeled female), 24 light colored two centimeter squares with a capital letter on each, 24 dark colored two centimeter squares with a small letter (same as the large letter) on each

Procedure:
1. Place half of the light colored squares in each container; do the same with the dark colored squares.
2. Shake the containers to mix the genes.
3. With your eyes closed, reach into the two containers at the same time and remove a gene from each.
4. Record the pair on your data table (use tic marks). Put the genes back into the original containers.
5. Each partner in the group should repeat steps 2 through 4 twenty-five times.

Results:

Genotype	# of Time Occurred	Phenotype

1. On your chart, label the phenotype for each genotype. Which phenotype occurred the largest number of times?

2. What is the dominant phenotype?

3. What is the recessive phenotype?

4. When you divide the dominant phenotype by the recessive phenotype, what ratio do you get?

5. What is the most common genotype?

6. What is the least common genotype? Is there more than one "least common"? If so, what is it?

7. What are the ratios among double dominant (2 capital letters), double recessive (2 small letters) and hybrid (1 capital and 1 small letter) genotypes?

8. What did each container represent?

9. What did each piece of paper represent?

10. What did each pair of papers represent?

11. What controlled which pieces of paper you drew?

12. If you have Bb for brown eyes, and you marry someone with bb for blue eyes, how likely are you to have a child with blue eyes?

13. Can your child have BB for brown eyes? Why?

Where Did You Get It?

B	B	B	B	B	B	B	B	B	B
B	B	B	B	B	B	B	B	B	B
B	B	B	B	B	B	B	B	B	B
B	B	B	B	B	B	B	B	B	B
B	B	B	B	B	B	B	B	B	B
B	B	B	B	B	B	B	B	B	B
B	B	B	B	B	B	B	B	B	B
B	B	B	B	B	B	B	B	B	B
B	B	B	B	B	B	B	B	B	B
B	B	B	B	B	B	B	B	B	B
B	B	B	B	B	B	B	B	B	B
B	B	B	B	B	B	B	B	B	B
B	B	B	B	B	B	B	B	B	B
B	B	B	B	B	B	B	B	B	B
B	B	B	B	B	B	B	B	B	B

Where Did You Get It?

b	b	b	b	b	b	b	b	b	b
b	b	b	b	b	b	b	b	b	b
b	b	b	b	b	b	b	b	b	b
b	b	b	b	b	b	b	b	b	b
b	b	b	b	b	b	b	b	b	b
b	b	b	b	b	b	b	b	b	b
b	b	b	b	b	b	b	b	b	b
b	b	b	b	b	b	b	b	b	b
b	b	b	b	b	b	b	b	b	b
b	b	b	b	b	b	b	b	b	b
b	b	b	b	b	b	b	b	b	b
b	b	b	b	b	b	b	b	b	b
b	b	b	b	b	b	b	b	b	b
b	b	b	b	b	b	b	b	b	b
b	b	b	b	b	b	b	b	b	b

Whose Eyes?

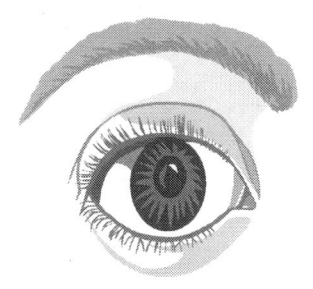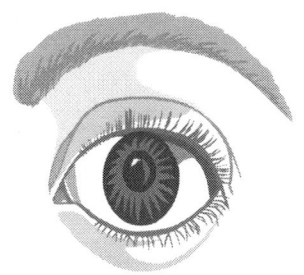

Objectives:
Students will demonstrate their dominant and recessive characteristics.
Students will determine their position on the genetic square.
Students will find the ratio of dominant to recessive traits for their class.

Materials:
Genetic square on transparency film (pattern attached) or transferred to a piece of poster board, water soluble markers or removable stickers for the poster board, overhead projector

Procedure:
1. Students will work in pairs, noting their dominant and recessive traits.
2. Use the chart to determine your traits. Circle your choice.

Characteristic	Dominant	Recessive
Attached ear lobe		aa
Free ear lobe	AA or Aa	
Widow's peak	WW or Ww	
Straight hair line		ww
Tongue rolling	RR or Rr	
No tongue rolling		rr
Inwardly bent little finger	BB or Bb	
Straight little finger		bb
Hairy fingers (2nd joint)	HH or Hh	
Bald fingers		hh
Perfect vision		vv
Imperfect vision	VV or Vv	
Five fingers/toes		ff
More fingers/toes	FF or Ff	

Results:
1. Use your data to determine where you are on the genetic square.
2. If two people end up at the same place on the genetic square, do they look alike? Why?

3. What would happen if you added another trait to the genetic square?

Whose Eyes . . .

4. Make a chart on the chalkboard with the number of people in your class under the correct traits. What is the ratio of dominant phenotypes to recessive phenotypes?

5. Is this close to the expected $^3/_4$ ratio? What could account for variations in the ratios?

6. Why is it important for scientists to study genetics?

Whose Eyes . . .

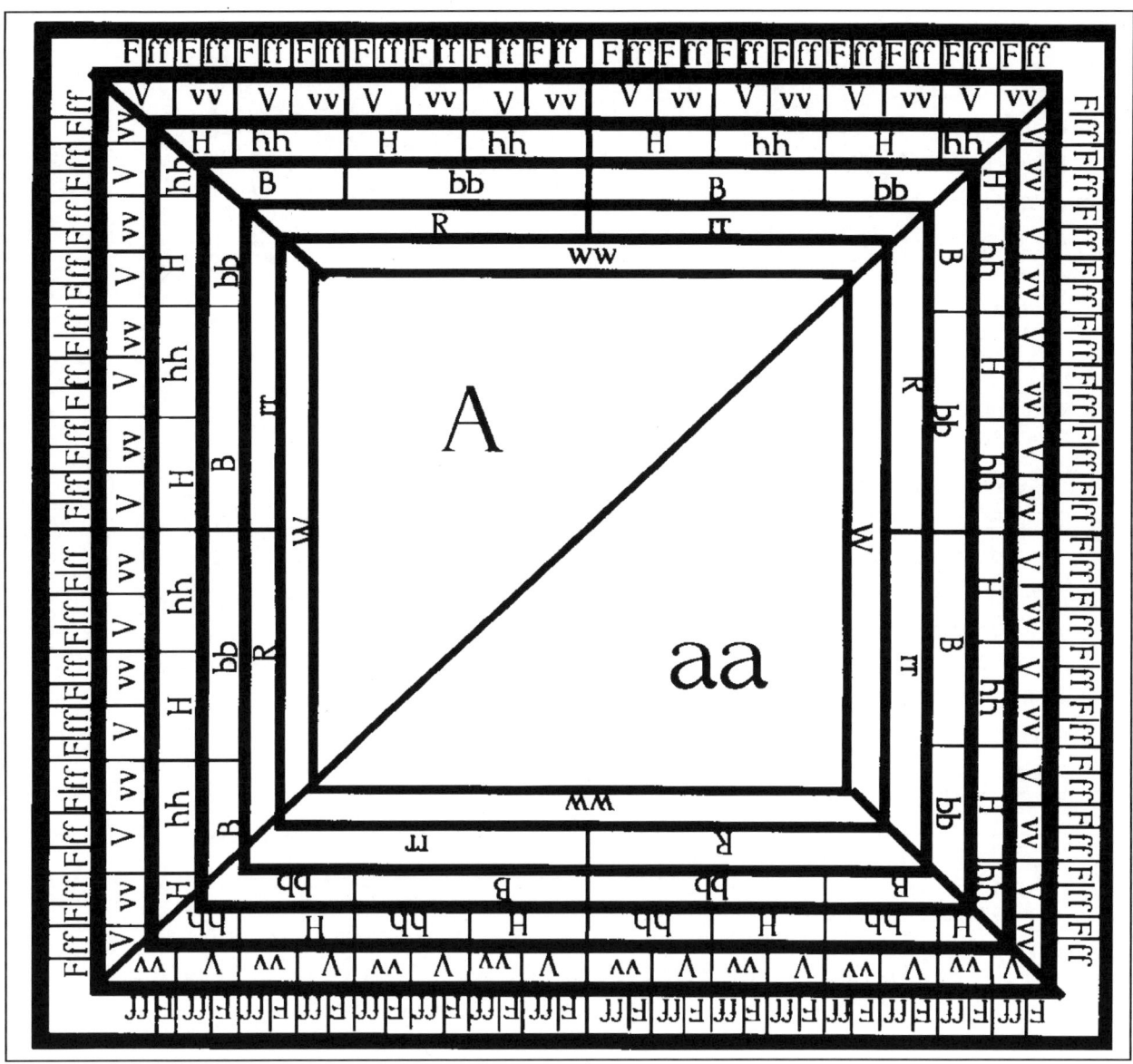

Wink or Blink

Objectives:
    Students will be able to identify differences between voluntary muscles and involuntary muscles.
    Students will give examples of both involuntary and voluntary muscles.
    Students will analyze data.

Materials:
    Watch with second hand

Procedure and Results:
1. Muscles may be classified into two groups: voluntary and involuntary. Voluntary muscles are those that you can control. Involuntary muscles are those that you cannot control. Some muscles such as those that control breathing and your eyelids can be controlled for a short period of time.
2. Blink your eyelids three times. Now try not to blink. Time your partner and yourself. Repeat two more times and record the times in the chart below.

Trial	Your Time	Partner's Time
1		
2		
3		

1. Are your eyelids really controllable? Why?

2. What other involuntary muscles can you control for a short while?

3. List examples of voluntary muscles.

4. Are there any involuntary muscles that you cannot control even for a little while? List them.

©2016 Near-Normal Design and Production Studio

Wonder Walk

Objectives:
   Students will identify living and non-living things in their environment.
   Students will observe the impact of humans on nature.

Materials:
   Coat hangers opened into a circle, plastic zip bags.

Teacher's Procedure and Results to Discuss:
1. Bend a coat hanger into a circle. Close the hanger portion to keep students from poking themselves or others.

2. Give each child or pair of children the coat hanger and a plastic bag.

3. Go outside on the schoolyard for a "treasure hunt." "Treasures" may include grass, rocks, pieces of paper, or anything they find on the ground they may want to look at more closely. Tell them they may not collect harmful items such as broken glass, needles, cigarette butts, and so forth.

4. Tell students to drop their coat hanger on the ground and hunt for "treasures" inside the hanger. You may want to allow them to hunt in more than one place.

5. Allow them to hunt for several minutes, collecting their "treasures" and placing them in the plastic bag.

6. Return to the classroom and have students place their "treasures" on a newspaper on the floor.

7. Have them divide the "treasures" into living and non-living piles.

8. Have students describe their findings and why they labeled them what they did.

9. Ask them to describe characteristics about living and non-living.

10. What do you do about objects discovered that once were living?

11. Which of their materials were left by humans? How does this impact the environment?

12. What can students do to protect their environment?

©2016 Near-Normal Design and Production Studio

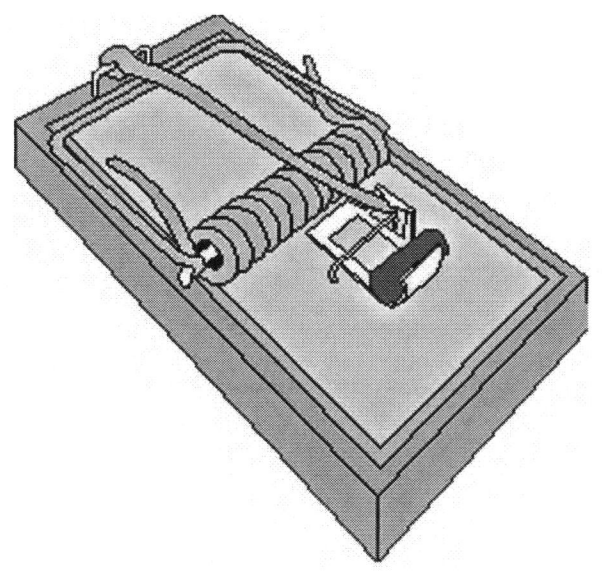

Physical Sciences

Balloons and Rice Krispies®

Objectives:
    Students will demonstrate static electricity.
    Students will observe discharge of electrical energy.
    Students will determine the relationship among the balloon, the puffed rice, and rubbing the balloon.

Materials:
    Balloon, some puffed rice, a piece of cloth or wool, balloon pump

Procedure and Results:
1. Put a few (up to ten) pieces of puffed rice in the balloon, then inflate it and tie the end.
2. Rub the balloon with the cloth as if you were trying to brush off some dust. Give the balloon 4 or 5 swift strokes; too many will lessen the effect.
3. Holding the top of the balloon with one hand, slowly move a finger from your other hand towards one of the puffed rice in the balloon. Describe what happened.

4. Try the same thing with several other pieces of the puffed rice and describe the results.

5. How can you explain these phenomena?

6. Repeat step #2 and bring your balloon close to your partner's balloon. Describe what happened to your partner's puffed rice.

7. Now both of you repeat step #2 and bring your balloons close together. Describe the results.

8. Why does the puffed rice respond as it does?

9. Why must you rub the balloon with the cloth?

©2016 Near-Normal Design and Production Studio

Boondoggle Derby!

Objectives:
   Students will observe the effects of forces on matter.
   Students will predict the outcome of an experiment from inferential data.
   Students will compare potential and kinetic energy.
   Students will describe the motion of the bottles.

Materials:
   Ramp (approximately 1 meter long), 3 plastic bottles (approximately 300ml each), 150ml water, 150ml glycerin (or a thick oil), 150ml steel or lead shot.

Procedure and Results:
1. Put 150ml water into one bottle, 150ml glycerin into the second, and 150ml shot into the third. Seal tightly.
2. Elevate one end of the ramp 15cm.
3. Predict which bottle, the water or glycerin, will get to the bottom of the ramp first.

4. Lay the bottles containing the water and the glycerin on the top of the ramp. Release them at the same time.

5. Which bottle got to the bottom of the ramp first?

6. Why do you think it got there first?

7. Take this bottle and the one containing the shot and place them on the ramp. Which do you think will make it to the bottom first? Why?

8. Release the bottles at the same time. What happened?

9. Why do you think this happened?

10. What general statement can you make about the contents of the bottles prepared for this race? Draw the forces at work on the bottles.

11. Stored energy is potential energy; kinetic energy is movement. How do potential and kinetic energy work in this experiment?

Bridge over the River "Wide"

Objectives:
    Students will design and carry out an experiment.
    Students will analyze weight distribution in bridge construction.
    Students will build a model of a bridge.

Materials:
    25 straws, 100 straight pins, a piece of cardboard (10cm x 45cm), weights, ruler.

Procedure and Results:
1. Each team must design and build a bridge using only the pins and straws. Keep in mind that the **pins are sharp**. The object is to construct the bridge that supports the most weight. The cardboard is used as a base for the bridge. **The bridge must span 40cm**.
2. You may want to draw sketches of your bridge for reference. All bridges must be finished before competition begins and no alterations are allowed during the contest. **Bridges must stand with weight for at least 1 minute**.
3. Complete the chart below.

Team	Weight Supported (in grams or kilograms)	Place (1st, 2nd, 3rd)

4. What reasons can you give for the winning entry's success?

5. How was the weight distributed on the winning entry?

6. If you could re-design your bridge, what would you change? Why?

©2016 Near-Normal Design and Production Studio

Deep Water

Objectives:
    Students will manipulate laboratory equipment.
    Students will predict the thickness of a film of water.
    Students will describe the behavior of water molecules.
    Students will describe the interactions of water with the environment.
    Students will calculate the thickness of a film of water.

Materials:
    Graduated cylinder, ruler (15cm), water supply, 1cm grid on transparency film and a flat, level surface

Procedures and Results:
1. Measure ten milliliters of water into the graduated cylinder.
2. Pour this slowly onto the grid on a flat, level surface (such as a table top).
3. Using your ruler, spread the water out into a rectangular shaped area, as large as you can make it without leaving holes in the water.
4. How does the water behave as you try to spread it?

5. How thick do you think the film of water is? Guess.

6. Measure and record the dimensions of your rectangle below.

    Length _____   Width _____

7. Calculate the area (Length x Width) _____. Check yourself by counting the number of squares on your grid the water covers.

8. How can you measure the thickness of the water? Remember, a milliliter is equal to 1 cubic centimeter.

9. Is there an indirect method of measuring this film's thickness? (Hint: V= L x W x H) Justify your answer.

10. From your experiment, how do you think the environment is affected by the way water behaves and by how thick the water molecules are?

11. How might we measure the thickness of a single sheet of notebook paper?

12. The present budget for the U.S. Government is more than three trillion dollars ($3,000,000,000,000). How many $1000 bills would this be?

13. If each bill were 0.004cm thick, how high would this stack of bills be? How did you determine this?

Dense, Who's Dense?

Objectives:
Students will collect information by measuring and recording data.
Students will operationally define density with relation to water.
Students will build a model of a hygrometer.

Materials:
Plastic straw, clay, BBs, sand, ruler, a waterproof marker, several small containers of unknown liquids, and a container of water

Procedure and Results:
1. Seal one end of the straw with clay.
2. Place enough BBs and sand into the straw to make it stand upright when placed in water.
3. With the marker, place a mark to indicate the level of the water on the straw. Be very accurate.
4. With a ruler, make three marks above and three marks below your first mark. The marks should be equally spaced, about 2 mm apart.
5. You have now constructed a hygrometer. Place it in one of the "unknown" liquids. Record the reading on the chart below. Repeat these steps for each of your liquids.

Liquid	Number of marks ABOVE	Number of marks BELOW
A		
B		
C		
Alcohol		
Oil		

6. Which liquids registered below the original mark?

7. Which ones registered above the mark?

8. Why did the hygrometer give you these different readings?

Dense, Who's Dense?

9. What does this information tell you about the oil? About the alcohol?

10. Did you get the same readings as the rest of the class? Why or why not?

11. Why did you use water to make your hygrometer? Why was the depth of your hygrometer in water the first mark you made on your hygrometer?

12. What does this lab have to do with density?

13. What is density?

Does Gravity Work?

Objectives:
    Students will observe the effects of cohesion.
    Students will operationally define air pressure.
    Students will predict the length of time the card will maintain its integrity.
    Students will relate air pressure to the environment.

Materials:
    Small plastic cup, index card large enough to cover the top of the cup, metric ruler, scale

Procedure:
1. Fill the cup to the top with water.
2. Place the index card over the top of the cup.
3. Holding your hand flat on the index card, invert the cup. **Keep your hand very flat.**
4. No water should drip out. Slowly remove your hand.

Results:
1. Why does the water stay in the cup?

2. How long will the card keep the water in the cup? How do you know?

3. Calculate the surface area of the index card (L x W) in square centimeters.

4. Find the mass of the water in grams.

5. Multiply the area of the index card by 1000 g/cm^2.

6. Which has more mass, the water pushing down on the card or the air pushing up on the card?

7. Do you still agree with your answer in question 1? Why?

8. How does air pressure affect your environment? Why do you think this?

Drip, Drip, Drip . . .

Objectives:
    Students will form and test their hypotheses.
    Students will experiment with surface tension.
    Students will operationally define variables.

Materials:
    Penny, dropper or pipette, paper towel, water supply

Procedure and Results:
1. Place the penny on a flat level surface such as a table top.
2. Guess how many drops of water you can place on the penny before it runs over onto the table. Your guess:

3. Slowly put drops of water on the penny and count them. How many did you get on the penny before it overflowed?

4. How did your experiment compare with your guess? Very Close _____ O.K. _____ Not Close _____

5. What observations did you make as you added drops of water to the penny? How do these observations explain your answer in question 3?

6. Compare your answer to question 3 with that of your classmates. What were the most drops and the least drops recorded?

    Most _____ Least _____

7. What factors (variables) do you think account for these differences?

8. Why do you think scientists are concerned about controlling variables when they do research?

9. How does this same reason apply to you in doing your classroom experiments?

Drip, Drip, Drip Challenge!

Objectives:
    Students will form and test their hypotheses.
    Students will design an experiment.
    Students will experiment with surface tension.
    Students will operationally define variables.

Materials:
    Penny, dropper or pipette, paper towel, water supply

Procedure:
    This challenge (if you choose to accept it) requires you to devise an experiment that will maximize the number of drops of water that can be placed on a penny.

Results:

1. What is your hypothesis?

2. Are there variables that you need to control? If so, what are they?

3. Are there some variables that you cannot control? If so, what are they?

4. Write down the design of your experiment.

5. What was your maximum number of drops?

6. What was the average of at least three tries?

7. Did you change your experimental design? If you did, how?

8. Did this change in design help or hurt your experiment? Explain.

9. Did you have to change your original hypothesis for the design of your experiment? Do you think scientists ever change their hypotheses? Why or why not?

Eureka!! I've Found It!

Objectives:
   Students will calculate the density of matter.
   Students will operationally define density.
   Students will apply experimental data to a problem.

Materials:
   Various solid objects, pan balance, container of water, metric ruler

Procedure and Results:
There was once a king who was given a beautiful golden crown. The king wanted to know if the crown was real gold, so he asked Archimedes to find out. However, Archimedes was not allowed to damage the crown by scraping any metal from it, nor could he use any chemicals on it. The solution came to Archimedes one day as he was getting into his bath. He yelled, "Eureka, I've found it!"

1. Measure the length, width and height of the small objects and record the data.

	Length	Width	Height	Area L x W	Volume L x W x H
Metal					
Wood					
Plastic					

2. Weigh the objects in the pan balance and record the mass below.

   Metal _____   Wood _____   Plastic _____

3. Compute the density of each object and record below. (D = V/M, D = density, V = volume, M = mass)

   Metal _____   Wood _____   Plastic _____

4. What happens to the object and to the water when you put each of these objects in water?

   Metal

   Wood

   Plastic

5. What does this have to do with how Archimedes determined the purity of the gold crown for the King?

Furlongs, Leagues, Stones and Pennyweights!!

Objectives:
- Students will measure his/her height using a variety of units.
- Students will describe a system of measuring.
- Students will graph heights.

Materials:
Meter sticks, yard sticks, variety of other objects such as tennis shoe, belt, tie, marking pens, butcher paper, scissors, and tape

Procedure and Results:
1. Explain to students how we could measure the length of a desktop in metric, English or "tennis shoes."
2. Have them help each other measure their heights in both metric and English.
3. Tape the butcher paper to the walls of the room. Arrange the students with their backs to the paper. Trace around their shoulders and heads. Label the tracing with the student's name and height in both English and metric units.
4. Who is the tallest? Who is the shortest? How tall is the tallest person in metric units? How tall is the shortest person in English units?

5. How could we make a graph of the class arranging the students from tallest to shortest? (With the teacher's permission, cut out your tracing and arrange it in order.)

6. Which measurement system did you use?

7. What did you do with the tracings of the students who were the same height?

8. Use your data to draw a graph plotting the frequency of students' heights. Attach your graph to this sheet.

9. The English system uses feet and inches to measure height; the metric system uses meters and centimeters. If you measured heights in "tennis shoes" what would the smaller units be?

10. Would those smaller units be as easy to use and as regular as centimeters? Explain.

11. Why is it important that we all agree on which measurement scale to use?

Go? No Go?

Objectives:
Students will identify variables associated with energy conversions.
Students will make a model of an electric motor.
Students will experiment with electricity.
Students will describe the system making up a motor.

Materials:
75cm Bell wire (#22 or #26 gauge), 1 small piece of emery board or sand paper, 2 large paperclips, 2 25cm pieces of insulated wire, 1 1.5v dry cell battery, 3 2cm x 2.5cm magnets, 2 small screws, 1 10cm x 20cm x 2.5cm piece of wood with small holes in two diagonal corners

Procedure and Results:
1. Roll the Bell wire around your index and middle finger leaving about 3cm sticking out on each end.
2. Loop the 3cm portion around the ends of the oval loop as illustrated.
3. This 3cm piece serves as the axle for the armature and must have the insulation removed from it. Scrape the enamel insulation with the emery board, being careful not to break the wire.
4. Bend the paperclips as illustrated and mount them to the wood.
5. Carefully remove the insulation from the tips of the other wire and connect them to the screws as illustrated.
6. Place the armature in the loops formed by the paper clips. Test their ability to turn freely and evenly.
7. Now place the magnets under the armature as illustrated. Connect the stripped tips of the wire to the battery and gently give the armature a spin to start. Experiment.
8. Does reversing the poles of the battery affect the motor? Explain.

9. Would the gauge of the Bell wire make a difference? Why?

10. Experiment! Will it run with two batteries? How did you hook them up?

11. Does it run faster or slower with two batteries? How about three or four? Experiment.

12. What are the variables you had to control in order for your motor to run?

The Great American Egg Drop

Objectives:
   Students will design and conduct an experiment.
   Students will interpret data from a chart.
   Students will analyze force placed on a structure.
   Students will apply the system they design to other structures.

Materials:
   Fresh eggs, 25 straws, and masking tape

Procedure and Results:
1. Design a container to support the egg and protect it from breaking when dropped. You may not modify the egg in any manner. What is your beginning hypothesis for the best design?

2. Each team must work independently and must finish its entry before any eggs are dropped. Once competition begins, entries cannot be modified. Describe your final design. Did your hypothesis change during the testing process, prior to completing the entry?

3. Use the chart to record data regarding the height from which the egg was dropped and survived unbroken.

Team	1 meter	2 meters	3 meters	4 meters	5 meters

4. What was the record height? What made the winning design the best?

©2016 Near-Normal Design and Production Studio

5. What forces did the egg have to survive? How were the forces distributed around the structure to keep the egg from breaking?

6. Distribution of forces around a body is important in the automotive industry. How does your experiment apply to this industry?

Hot Tamales

Objectives:
    Students will demonstrate chemical acid/base reactions.
    Students will observe chemical change.
    Students will demonstrate oxidation reactions.
    Students will apply chemical change to body processes.

Materials:
    Dirty penny, picante sauce and forceps

Procedure:
1. Make at least 5 observation of the penny and write these in the Results section.
2. Holding the penny with the forceps, place it in the picanté sauce.
3. At the end of 5 minutes, make 5 more observations the penny.

Results:
1. Observe the penny prior to putting it in the picanté sauce.

2. Observe the penny after removing it from the picanté sauce.

3. What could cause the changes you observed?

4. What do you think the picanté sauce has in it that could cause this change?

5. Is the change in the penny chemical or physical? Support your answer.

6. Rust is the result of the oxidation of iron. Do you think the penny has undergone oxidation? Explain your answer.

7. How are the changes caused by the picanté like the changes food undergoes in the digestive system?

8. Other than to eat, what else could you use picanté sauce for?

Leaky Pipes

Objectives:
  Students will correlate the rate of water loss from a leak with the cost of the water.
  Students will calculate the rate of water loss.
  Students will apply information to environmental interactions.

Materials:
  Water; 50ml, 100ml and 200ml graduated cylinders; funnel; faucet; and a clock/watch with a second hand

Procedure and Results:
1. Place the funnel into the 50ml graduate. Turn on the water at a very slow leak. When the drip is steady, place the funnel and graduate under the drip for 1 minute. Measure and record the amount of water in the graduate.
2. Repeat this with the 100ml and 200ml graduates at even faster rates. Record the measure of volume of each.
3. Assume 1 liter costs 1 dollar; calculate the loss in dollars for each graduate.

4. Calculate the water loss in dollars and liters that would occur in one day at each rate of drip.

5. Calculate the water loss in dollars and liters that would occur in one year at each rate of drip.

6. Complete the following:
   (a) What is the significance of faulty plumbing?

   (b) How can you decrease the amount of water used by a toilet but not change its efficiency?

   (c) Why is it necessary to measure the amount of water that is wasted?

Leaning Tower of Pizza!

Objectives:
  Students will design and conduct an experiment.
  Students will analyze forces on a structure.
  Students will build a model of a tower.

Materials:
  25 straws, masking tape, one electric fan, meter tape

Procedure and Results:
1. Design and construct a tower capable of withstanding a "hurricane wind" (supplied by the fan). Each team must work independently and must be finished before competition begins. No alterations are allowed once the contest has begun. What is your hypothesis for the design of your tower?

Explain your design.

2. For each tower, move the fan towards the tower until it blows over. Record this critical distance. The winner is the entry which gets closest to the fan and has the greatest height (25 straws laid down on the floor will not work). You may wish to make sketches of each of your competitors' towers and of your tower for reference.

3. Complete the chart and label the top three entries.

Team	Height	Distance to Fan

4. What reasons made the best tower the winner?

5. From what directions were the forces on the tower?

6. How could the winning design be applied to the housing industry?

Magnets and "Attractive" Forces

Objectives:
   Students will demonstrate magnetic polarity.
   Students will apply data to the earth's pole.

Materials:
   A pair of magnets, a sheet of paper, some iron filings (Do not drop or strike the magnets.)

Procedure and Results:
1. Play with the magnets for a moment and draw what you observe in the space below.

2. Place the magnets in front of you with opposite poles close together but not touching.
3. Cover the magnets with the paper.
4. Sprinkle the filings on top of the paper; draw what you observe.

5. Rearrange the magnets with like poles near but not touching, replace the paper, and sprinkle the filings again; draw what you observe.

6. What conclusions can you make from your observations?

7. Why do the filings line up in the manner they do?

8. What are the filings "stuck" to?

9. How are the poles of the magnets like the poles of the earth?

10. How are the lines of force from the magnets like the magnetic fields around the earth?

Match Me if You Can!!

Objectives:
    Students will compare sounds and match those that are alike.
    Students will make observations.
    Students will determine that sound waves need energy to travel.

Materials:
    10 film cans (available through Amazon), five numbered 1-5, five lettered a-e, objects such as an eraser, a paper clip, BBs, sand, water, etc.

Procedure and Results to Discuss:
1. Give the children a set of five containers, reminding them not to touch or open them until directed.
2. Select one of the cans and shake it several times making sure the children are listening.
3. Ask the children to shake each of his/her containers and try to match the sound.
4. Repeat with all containers.
5. Ask the children to describe the sounds they heard with in words (rolly, clanky, swishy, etc.)
6. Why is it important to make observations with more than just our eyes?

7. What do you have to do to make sounds come from the cans?

Mirror Images, segamI rorriM

Objectives:
   Students will classify variables.
   Students will operationally define mirror images.
   Students will observe light energy.

Materials:
   Paper, pencil, two small mirrors, and ruler

Procedure and Results:

1. Look at yourself in the mirror. What do you notice?

2. Can you see your entire face? Why or why not?

3. How far away does your face appear to be?

4. Hold the mirrors so that they form a 90° angle with each other. Make sure the image of your nose is exactly where the two mirrors meet. Hold the mirrors far enough away to see your left ear. Touch your left ear. Is it on the left in the mirror? Describe what has happened.

5. Place the sheet of paper on a desk surface. Place the mirror perpendicular to the paper so that you can see the lines of the paper in the mirror. You may want to cover your hand with another piece of paper. Looking only in the mirror, **not at your hand or the paper**, print your name on the paper. How did it go?

6. Keep trying until your name is written correctly. Now look at the paper directly. How does it read? Why did this happen?

7. Draw a line about 2.5cm long on your paper. Hold the two mirrors so that you can see the line. Slowly begin bringing the two mirrors together. What happens to the image of the line?

8. What is meant by the term "mirror image"?

9. Give 3 examples of mirror images.

10. List three variables you changed throughout this activity. What responded to those changes? What variables did you keep the same?

11. Is there energy being used when you see an image in a mirror? Explain your answer.

Mirror, Mirror on the Wall

Objectives:
    Students will describe similarities and differences between objects.
    Students will operationally define mirror images.
    Students will observe light energy.
    Students will describe differences between reflected and real objects.

Materials:
    Paper, pencil, and small mirror

Procedure and Results:

1. Look at yourself in the mirror. What do you notice about the size of the image?

2. Can you see your entire face? Can you see your entire body?

3. How far away does your face appear to be?

4. Touch your left ear lobe. Is it on the left in the mirror?

5. Place a sheet of paper on a desk surface. Place the mirror perpendicular to the paper so that you can see the lines of the paper in the mirror. Looking only at the mirror, print your name on the paper. You may want to have your partner hold a sheet of paper between you and your hand so that you can only see your writing in the mirror. How did it go?

6. Keep trying until it is printed correctly. Now look directly at the paper. What can you say about your name?

7. Slowly tilt the mirror toward the written name. What happens?

8. What is meant be the term "mirror image"?

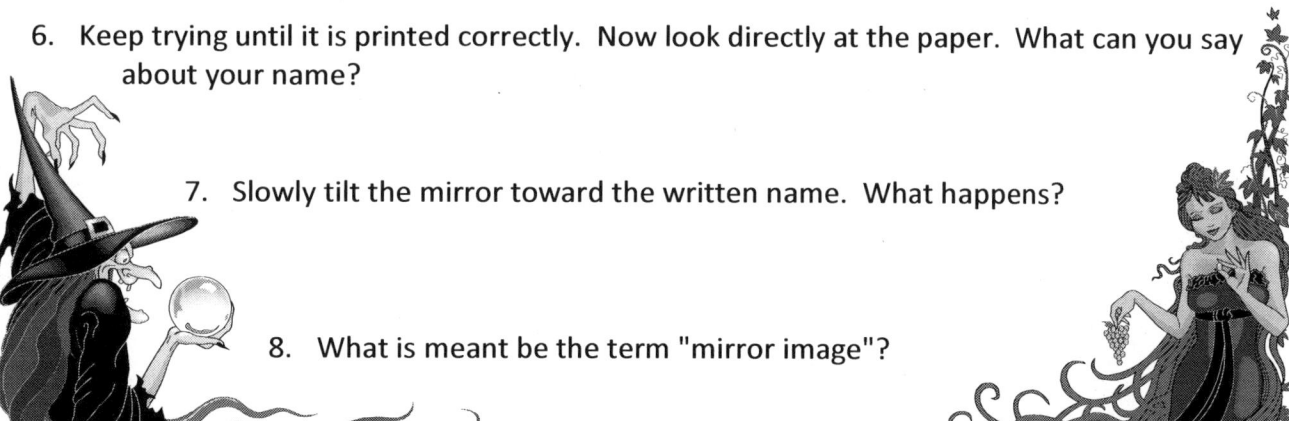

9. List some examples of "mirror images".

10. How is what you see in the mirror different from what is real? How is what you see like what is real?

11. Do you need light to use a mirror? How do you know?

Screwdrivers, Big and Small

Objectives:
    Students will determine cause and effect relationships.
    Students will operationally define mechanical advantage.
    Students will calculate mechanical advantage.

Materials: Screwdrivers of various sizes, screws, and boards with holes in them

Procedure and Results:
1. Use screwdriver #1 to put one of the screws into the board. Do the same with screwdrivers #2 and #3.
2. Which screwdriver was the easiest to turn?

3. Which screwdriver was the hardest to turn?

4. What are some similarities among the three screwdrivers? What are some differences?

5. What accounts for the ease of turning the screwdrivers?

6. Calculate the mechanical advantage (MA) of each of the screwdrivers. Remember that MA = Dw/Da (Dw = diameter of the wheel; Da = diameter of the axle).

    MA (#1) = _____

    MA (#2) = _____

    MA (#3) = _____

7. If you were going to design a screwdriver, what guidelines would you follow? What characteristics would make your design effective?

8. Look at the two illustrations below. Which design would you choose for a:
   (a) gain in speed? _____

   (b) gain in power? _____

9. How is this activity related to bicycles?

 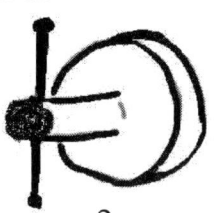

1   2

Solid to Liquid, Ice to Water

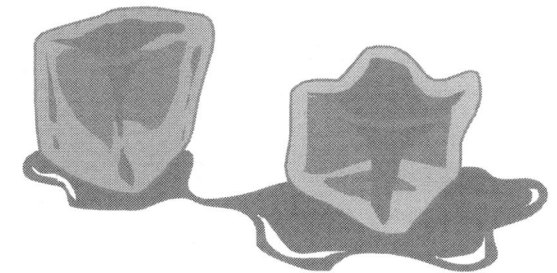

Objectives:
    Students will accurately read a thermometer.
    Students will observe physical phenomena.
    Students will observe a change of state.
    Students will observe energy of heat.

Materials:
    Ice, thermometer and a container of water (**Do not consume any of the materials.**)

Procedure and Results:

1. Observe the ice in the glass during the class time today. Record your visual observations.

2. Using a thermometer, check and record the temperature of the water every ten minutes during the class.

	Before Ice	After Ice	10 minutes	20 minutes	30 minutes	40 minutes
Temperature						

3. What factors affected the initial temperature?

4. What factors affected the last temperature reading?

5. What seems to be happening to cause the ice to melt?

6. What is happening to the energy in the glass of water? What is the source of the energy?

7. HOMEWORK ASSIGNMENT: With the help of an adult, boil some water and record your observations. You'll need to measure the temperature as you did in class.

©2016 Near-Normal Design and Production Studio

Solutions and Mixtures

Objectives:
 Students will identify properties of solutions and mixtures.
 Students will demonstrate separation of materials.

Materials: Small test tubes, water, vinegar, alcohol, beakers, filter paper, water soluble black ink, scissors, and cooking oil

Procedures and Results:
1. Label three test tubes "W", "A", and "V", and fill them with 5ml of water, alcohol and vinegar respectively.
2. To each test tube, add a pinch of salt. Placing your thumb over the top, shake each vigorously. Describe your results for each. What happened to the salt? Why?

	Salt Observations	Explanation
Water		
Alcohol		
Vinegar		

3. Pour out the liquids, rinse the test tubes, and refill them with fresh water, alcohol and vinegar. Make sure the liquids are in the correctly labeled test tubes. Add fifteen drops of oil to each test tube and shake again. Record your observations. What happened to the oil? Why?

	Oil Observations	Explanation
Water		
Alcohol		
Vinegar		

4. Cut the filter paper into a strip so that it will hang down into a beaker; the bottom of the paper should not touch the bottom of the beaker. Put a drop of the black ink about 2cm from the tip of the strip and put enough water in the beaker so that the paper just touches the water. Record your observations.

5. Did any of these experiments show chemical or physical changes? Can these processes be reversed? Explain.

6. Which of all the above experiments were mixtures? Which were solutions? Which were emulsions? Which were tinctures?

7. Give an example of a mixture that you can eat.

Sound, Noise, and Bottle Music

Objectives:
   Students will observe differences in musical sounds.
   Students will identify factors affecting musical tones.
   Students will demonstrate the movement of sound waves.
   Students will determine the relationship between air columns and pitch.
   Students will analyze factors necessary to produce sound.

Materials:
   Several bottles of various sizes and shapes, water, and a metal object (pencil, spoon, etc.)

Procedure and Results:
1. Fill each bottle with water to the same level (about ¼ full). Tap each with the metal object. What do you hear?

2. Although the bottles are filled to about the same level, is there a difference in the sound? Explain.

3. Fill each bottle half-full of water. Tap each again. What do you hear? How is it different from what you heard before?

4. What conclusions can be drawn from these phenomena?

5. What other objects can be "tuned" to produce different sounds or notes?

6. With this in mind, what is the difference between noise and music?

7. To produce a sound, something has to vibrate. What vibrated in this experiment?

8. Could you hear the notes you played on the bottles if you were in a vacuum? How do you know?

Standard Weight... A Potato!

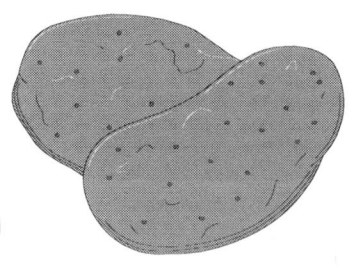

Objectives:
    Students will manipulate lab equipment.
    Students will classify objects or events according to similarities and differences.

Materials:
    Classroom articles, (textbook, ruler, scissors, eraser, ball, etc.) spring-balance, paper, pencil, and a potato

Procedure and Results:
1. Use your potato as a "standard" of mass and estimate the order of your classroom articles (from least to greatest weight). Record your estimates in the chart below.
2. Weigh each of the articles and record its relative order and its weight.

Article	Estimated Order	Weighed Order	Weight	Difference (+ or -)
Potato				0 Newtons

3. What are the advantages of the potato as a "standard" of mass?

4. Weight of the potato _____ Find the difference between each article and the potato.

5. What are the disadvantages of using such a standard?

6. What standards do we use in the United States for determining weight?

7. What are the advantages/disadvantages of our standards?

8. What other standards are used?

9. What are the advantages of using an agreed upon standard?

10. How is measuring weight a form of classification?

Static and "Electric Doorknobs"

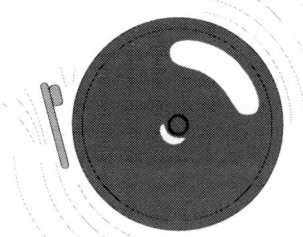

Objectives:
  Students will demonstrate static electricity.
  Students will identify different static electricity charges.

Materials:
  Comb, rubber rod, plastic rod, pith balls, string, ring stand, test tube, clamp, wool or nylon cloth and paper.

Procedure and Results:
1. Tear a small bit of paper into tiny bits. Hold the comb over the bits and describe the results.

2. Now comb through your hair a few times and hold the comb over the paper bits again. What happened?

3. Hang the pith ball from the clamp with the string, comb your hair again and move the comb close to the ball.
   What happened?

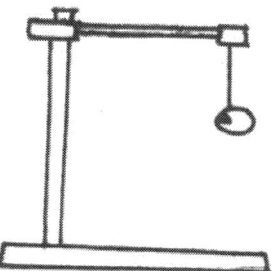

4. Rub one of the rods on some soft cloth a few times. Move the tip of the rod close to the ball and record the results.

5. Rub the rod on some thick animal fur and move the tip towards the ball. Describe the results.

6. Make a drawing to indicate what is happening electrically in this experiment. Use a plus (+) sign to show a positive charge and a minus (-) sign for a negative charge.

7. What did you have to do to generate static electricity? Why did you have to perform this action?

8. On dry days you can walk across a carpet, touch a doorknob and get an electrical shock. Why does this happen?

Stretch, Bend, Rub or Warp

Objectives:
　Students will identify properties of various objects.
　Students will make observations.

Materials:
　Rubber bands, paperclips, aluminum foil, craft sticks, metric ruler and scissors

Procedure and Results:
1. Stretch the rubber band back and forth very quickly. What does it feel like when you finish? Did you feel a difference in temperature immediately after you stretched it?

2. What happened to the rubber band when you pulled it out and let go? Stretch the rubber band a few more times. Observe the distances you can stretch it without breaking it and the amount of recovery.

3. What is this property of the rubber band called?

4. Bend the paperclip back and forth several times slowly. What happens?

5. Bend another paperclip back and forth rapidly. What happens?

6. What property does the paperclip exhibit?

7. Cut a small piece of the aluminum foil into a square and measure each side. Using your craft stick, smooth the foil working from the center out. Have your teacher demonstrate this technique. What are the beginning dimensions?

©2016 Near-Normal Design and Production Studio

Stretch, Bend, Rub or Warp

8. What are your dimensions after you have smoothed it with the craft stick?

9. What can you say about what happened to the foil?

10. What are some of the properties of matter you learned about today? Do you think all matter has these properties? Why or why not?

Up, Down, All Around!

Objectives:
  Students will identify motions.
  Students will make observations.
  Students will make predictions.

Materials:
  A mechanical toy (dog, cat, robot, etc.)

Procedure and Results to Discuss:
1. Show a toy to the class.

2. Ask children how it will move. Accept as many answers as the children can generate.

3. Set the toy in motion. Have the children imitate the movements.

4. Ask the children for the directions of the movements (up, down, around, etc.)

5. What other types of motion do the children see at school?

6. Bring out a toy they have not seen before. Ask the children to guess what type of motion it will have. Set the toy in motion and let children describe its movements.

What Is Full?

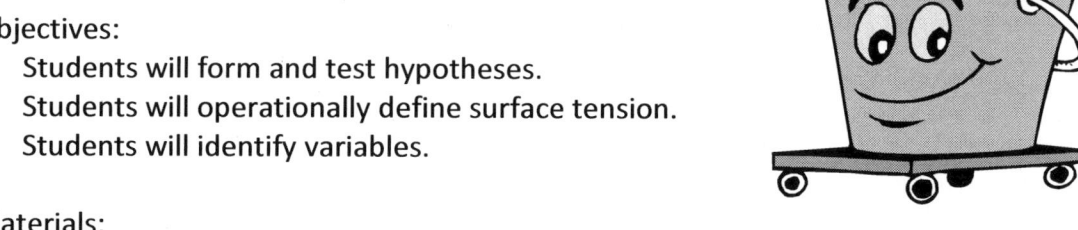

Objectives:
   Students will form and test hypotheses.
   Students will operationally define surface tension.
   Students will identify variables.

Materials:
   Small plastic container (2 ounce condiment cup), water, pennies, tweezers or clothespin

Procedure and Results:

1. Completely fill the container with water. Draw a side view of the container, accurately marking the water level.

2. How many pennies will you be able to add to the container before the water overflows?

3. Carefully add pennies, one at a time, to the water. Try to disturb the surface of the water as little as possible. (Use the clothespin or tweezers.) How many did you add before the water overflowed?

4. How accurate was your guess in question # 2?

   _____Very Close _____O.K. _____Not Close

5. Describe what you saw as you added pennies to the water. You may want to draw the position of the water just before it overflowed.

6. Remove and dry the pennies. Repeat steps 1-3 but add some soap to your water. What happened this time?

7. What did the soap do to the water?

8. Water has a property called surface tension. From your experiments, what does surface tension mean?

9. What did you change in the second part of this experiment (step 6)? What responded to your change?

Index

adaptation, 101, 140
air
  currents, 55
  pollution, 103
  pressure, 31, 167
  weather, 31
analyze, 162
analyze data, 109, 136, 154

bridge, 162

calculate, 163, 173, 181, 191
change
  chemical, 179
  state, 193
circulatory system, 106, 107
  heart rate, 107
classify, 14, 19, 21, 25, 65, 73, 105, 111, 125, 187, 198
classify, 11
climate, 34
cohesion, 167
compare, 63, 134
conduct research, 101
construct, 123
construct, 63, 64, 67, 86, 128, 162, 182
Coriolis effect, 82
correlate, 181
create an experiment, 113

density, 165, 173
describe, 11, 14, 19, 25, 45, 55, 83, 101, 155, 160, 163, 174, 176, 186, 189
design, 17
determine, 17, 22, 31, 34, 38, 45, 57, 59, 63, 69, 79, 96, 103, 134, 140, 142, 144, 146, 151, 159, 186, 191, 196
determine relationships, 34, 36, 59, 71, 79, 92, 159, 196
dinosaur, 86
Doppler Effect, 92
draw conclusions, 101, 136

earth
  craters, 77
  magnetic fields, 184
  materials, 73
  rotation, 69, 79
electric motor, 176
electricity, 176
  charges, 200
  static, 159, 200
energy
  conservation, 96
  conversion, 176
  electricity, 159
  heat, 193
  insulation, 63
  kinetic, 160
  light, 187, 189
  non-renewable, 96
  potential, 160
  radioactivity, 71
  recycle, 132
  renewable, 96
  sound, 186
  types, 96
environment
  abiotic, 36, 115, 123
  air pressure, 167
  biotic, 36, 115, 123
  energy, 38, 57
  insulation, 63
  interactions, 128, 140
  man's impact, 61, 155
  ocean, 83
  process, 21
  solid waste, 132
  water conservation, 181
erosion
  splash, 77
estimate, 55, 79, 132, 198
ethics, 38
examine, 67
experiment, 162, 171, 177, 182
explain, 11, 14, 17, 21, 31, 105, 130

force, 45, 160, 177, 182

genetics
  chromosomes, 125
  crossbreed, 130
  dominant, 111, 146, 151
  genotype, 130
  genotypes, 146
  phenotype, 130
  phenotypes, 146
  recessive, 111, 146, 151
graph, 57, 63, 71, 174
greenhouse effect, 36

habitat, 123
half-life, 71
health
  diet, 144
  nutrition, 144
  smoking, 138
heart rates, 107
hypothesize, 34, 38, 45, 65, 86, 96, 118, 142, 169, 171, 205

identify, 22, 33, 61, 86, 101, 106, 107, 154, 136, 194, 204
infer, 81
interpret, 38, 71, 128t, 144, 177

living, 105, 155

magnet, 33, 184
magnetism, 33
map, 83
matter, 160, 173
measure, 34, 59, 69, 79, 113, 120, 132, 165, 174
measure,
   temperature, 34
measure
   pulse, 106, 107
measure
   pulse, 107
mechanical advantage, 191
microbes, 118
mineral, 73
mirror images, 187, 189
mixture, 194
model, 45, 47, 51, 61, 125, 128, 162, 165, 176, 182
moon
   craters, 77
   phases, 65
motion, 160, 204
muscular system
   involuntary, 109, 154
   voluntary, 109, 154
nervous system, 134
   eye, 134
   reaction, 121
   reflex, 121
non-living, 105, 155
observe, 26, 45, 55, 73, 81, 106, 107, 113, 115, 118, 121, 125, 134, 138, 155, 159, 160, 167, 179, 186, 187, 189, 193, 196, 202, 204
ocean
   currents, 83
operationally define, 36, 57, 71, 92, 111, 165, 167, 169, 171, 173, 187, 189, 191, 205

Pepper Moth, 101
pie chart, 38
pollution
   air, 103
   water, 142
predict, 52, 57, 93, 95, 107, 123, 140, 146, 160, 163, 167, 204
properties
   physical, 202

ratio, 151
reaction
   chemical, 179
reactions
   acid/base, 179
relative age
   craters, 77
resources
   non-renewable, 52, 57, 61
   renewable, 52, 57, 61, 93, 95
   water, 93
rock, 73
rock layers, 45

safety, 22, 65
sense of touch, 19
shapes, 25
solution, 194
sound, 186, 196
sound waves
   Doppler Effect, 92
speed
   Doppler Effect, 92
star, 67
star charts, 67
sun
   rise, 69, 79
surface tension, 169, 171, 205

temperature, 34
tobacco, 138

variable
   manipulated, 142
   responding, 142
variables, 17, 113, 169, 171, 176, 187, 205
   light, 113
   water, 113, 142

water
   conservation, 142
   molecules, 163
weather, 31, 36, 55
   clouds, 59
   dew point, 59
   wind, 81

Made in the USA
Lexington, KY
04 May 2018